Airport Technology Research Plan
...for the NextGen Decade

2012 - 2021
A $250M Investment in Airport Safety
January 2012

U.S. Department of Transportation
Federal Aviation Administration

FAA William J. Hughes Technical Center
Atlantic City International Airport,
New Jersey 08405

Airport Technology Research Plan
...for the NextGen Decade

This booklet describes the serious contemporary issues facing airport technology and outlines an essential Federal Aviation Administration (FAA) research and development plan to deal with these issues. With the implementation of new procedures from the NextGen research, the role of airports will be to accommodate the increased traffic safely. This is especially critical during aircraft operations in inclement weather. This will require development of technologies to heat airport pavements, reliable methods to assess the braking performance of aircraft, development of lighting and marking materials providing higher visibility, development of new lighting technologies, such as, holograms, developing methods to mitigate wildlife at or near the airport, and developing new and efficient techniques for aircraft rescue and fire fighting.

The latest airport pavement design and evaluation advisory circular, AC 150/5320-6E, which includes the design procedure FAARFIELD, is considered to be suitable for thickness design for the current fleet of heavy aircraft with a high degree of confidence in the thickness computations as affected by gear configurations and wheel loads. This software tool is recognized by the international aviation community for worldwide use. Other aspects of pavement design, construction, and evaluation, such as mix design, material selection, environmental factors, maintenance, sustainability, and life-cycle description, need to be studied and procedures developed in the light of changing governmental responsibilities and fiscal regimes. In addition, we will evaluate the feasibility of extending the design life of runway pavements at major hub airports from 20 years to 40 years, with, presumably, the expectation that life extension would be required for other airport pavements if it is shown that the concept is feasible and practically achievable. We will study these aspects of airport pavement operations, starting with a discussion of how to define pavement life under these new conditions and why a more precise definition of life-cycle cost analysis is required to satisfactorily meet evolving requirements over extended periods of time.

Successfully executing the enclosed plan will reaffirm FAA's leadership responsibilities as a key player of the aerospace industry.

For more information please contact:

Aviation Research Division
Airport Technology R&D Branch
Federal Aviation Administration Technical Center
Atlantic City International Airport, NJ 08405

Attention: Dr. Satish K. Agrawal
Telephone: (609) 485-6686

Table of Contents

EXECUTIVE SUMMARY	2
Airport Safety Technology Research And Development	
1.0 Introduction	4
2.0 Research Direction	5
2.1 Winter Operations	5
2.2 Braking Performance Reporting	6
2.3 Aircraft Rescue and Firefighting	6
2.4 Visual Guidance (LED, Runway Incursions, Nanotechnology, and Holograms)	7
2.5 Wildlife Hazard Mitigation	7
3.0 Winter Operations on Airports: Heated Pavements	9
4.0 Winter Operations on Airports: Aircraft Braking Performance on Ice and Snow	11
5.0 Aircraft Rescue and Firefighting: Firefighting Operations for Cargo Aircraft	13
6.0 Next Generation Aircraft Rescue and Firefighting	15
6.1 Advanced Composite Materials Firefighting	15
6.2 Biofuels	15
7.0 Aircraft Rescue and Firefighting: Second Level Access for NLA	17
8.0 Visual Guidance: Safety, Capacity, and Environment	18
9.0 Wildlife Hazard Mitigation	20
10.0 Airport Planning and Design	23
10.1 Airport Planning	23
10.2 Aircraft Noise and Sleep Annoyance Research	25
11.0 Airport Technology Research Taxiway	26
Airport Pavement Technology Research And Development	
1.0 Introduction	29
2.0 Airport Pavement Life	32
3.0 FAA PAVEAIR and Other Support Software	38
3.1 FAARFIELD	40
3.2 COMFAA 2.0 and 3.0	42
3.3 BAKFAA	42
3.4 ProFAA	42
3.5 Software Integration	43
4.0 Life Cycle Cost Analysis	44
4.1 Materials and Construction Costs	45
4.2 Sustainability	46
4.3 Integration with the Thickness Design Procedures	48
4.4 LCCA Software Integration with PAVEAIR	50
5.0 Non-Destructive Testing (NDT)	50
5.1 Pavement Profiling and Groove Measurement	51
6.0 New Materials and Processes	51
7.0 National Airport Pavement Test Facility (NAPTF)	53
8.0 Field Testing	55
9.0 High Temperature Pavement Test Facility (HTPTF)	56
References	58
Milestone Charts and Costs	60

EXECUTIVE SUMMARY

The fundamental mission of the Federal Aviation Administration is to foster a safe and efficient air transportation system. Our air transportation system has over 19,700 landing facilities, 246,000 registered aircraft, 594,000 pilots, 15,000 tower controllers, a multitude of terminal buildings and access roads, and 706 million passenger enplanements each year. In this massive system, there are many opportunities for human errors which can lead to accidents and incidents.

Current trends indicate that the aircraft fleet will increase not only in number, but more importantly, they will become more diverse. Very light jets, general aviation and air carrier aircraft constructed in part of composites, and new large aircraft with second level passenger seating will greatly increase in number. Over the next ten years, the number of enplanements is expected to eventually pass the 1 billion mark.

Our goal is to accommodate the projected traffic growth and establish an operational environment which is free of accidents or fatalities. Since further expansion of our airport facilities is limited by economic and environmental constraints, we must continuously improve the system and maximize the use of existing facilities by developing new standards, criteria, and guidelines to plan, design, construct, operate, and maintain the Nations's airports.

The report lays out a 10-year research plan with a $250 M investment in airport safety and capacity. It investigates the technologies, methods, materials, and procedures that will be required to accommodate the traffic growth as a result of implementation of NextGen technologies in our airport system.

The Airport Technology Research Program supports the agency's mission by conducting the necessary research and development required to enhance the safety of operations at our nation's airports and to ensure the adequacy of engineering specifications and standards in all areas of the airport systems and, where necessary, develop data to support new standards. With the implementation of new procedures from the NextGen research, the role of airports will be to accommodate the increased traffic safely. This is especially critical during aircraft operations in inclement weather. The increased traffic will necessitate efficient inspection and maintenance of our runways and taxiways. This will require development of technologies to heat airport pavements, reliable methods to assess the braking performance of aircraft, development of lighting and marking materials providing higher visibility, development of new lighting technologies, such as, holograms, developing methods to mitigate wildlife at or near the airport, and developing new and efficient techniques for aircraft rescue and fire fighting.

We have made significant progress in the development of the airport pavement design procedures for accommodating the new large heavy air transport, such as, Boeing B-777 and Airbus A-380 aircraft. The construction of the National Airport Pavement Test Facility has provided the FAA a unique testing capability which has yielded the computer software program FAARFIELD. This software is included in the latest airport pavement design and evaluation advisory circular, AC 150/5320-6E, which is considered to be suitable for thickness design for the current fleet of heavy aircraft with a high degree of confidence in the thickness computations as affected by gear configurations and wheel loads. This software tool is recognized by the international aviation community for worldwide use. Supplementary design and evaluation computer programs were also developed, including the pavement management system FAA PAVEAIR.

Other aspects of pavement design, construction, and evaluation, such as mix design, material selection, environmental factors, maintenance, sustainability, and life-cycle description, need to be studied and procedures developed in the light of changing governmental responsibilities and fiscal regimes. In addition, we will evaluate the feasibility of extending the design life of runway pavements at major hub airports from 20 years to 40 years, with, presumably, the expectation that life extension would be required for other airport pavements if it is shown that the concept is feasible and practically achievable. We will study these aspects of airport pavement operations, starting with a discussion of how to define pavement life under these new conditions and why a more precise definition of life-cycle cost analysis is required to satisfactorily meet evolving requirements over extended periods of time.

Airport Safety Technology Research and Development

1.0 Introduction

The FAA's continuing mission is to provide the safest, most efficient aerospace system in the world. Our air transportation system has over 19,700 landing facilities, 246,000 registered aircraft, 594,000 pilots, and 15,000 tower controllers, a multitude of terminal buildings and access roads, and 706 million passenger enplanements each year. In this massive system, there are many opportunities for human errors which can lead to accidents and incidents.

Current trends indicate that the aircraft fleet will increase not only in number, but more importantly, they will become more diverse. Very light jets, general aviation and air carrier aircraft constructed in part of composites, and new large aircraft with second level passenger seating will greatly increase in number. Over the next ten years, the number of enplanements is expected to eventually pass the 1 billion mark.

The Airport Safety Technology Research Program supports the agency's mission by conducting the necessary research and development required to enhance the safety of operations at our nation's airports and to ensure the adequacy of engineering specifications and standards in all areas of the airport systems and, where necessary, develop data to support new standards. In searching for solutions to accommodate increased "service demand" consistent with safety standards, the role of new technology is to strike a balance between the safety and efficiency of the "system" on one hand and the expenditure of public funds on the other hand. That balance is also subjected to environmental and economic constraints which will play an ever increasing role in the development of new technology in the airport systems research.

The research conducted within the Airport Safety Technology Research Program directly supports the FAA's Advisory Circular system, which is the principal means by which the FAA communicates with the nation's airport planners, designers, operators, and equipment manufacturers. These Advisory Circulars, commonly referred to as an AC, present the standards used in the design, construction, installation, maintenance, and operation of airports and airport equipment. Additionally, the AC provides current advice on airport operational and safety topics. To date, the research conducted within the Airport Safety Technology Research Program has provided the necessary technical data to support hundreds of ACs that have been published on a wide range of technical subjects. These technical subjects include airport design standards, visual guidance aids such as lighting marking, or navigational aids, airport rescues and firefighting equipment and procedures, pavement surface conditions, wildlife mitigation and detection, airport capacity enhancements, pavement friction, and snow and ice mitigation.

As the FAA pursues it's wide ranging transformation of the entire national air transportation system with the new NextGen program, it is important that the research conducted under the Airport Safety Technology R&D program be properly aligned to meet the future demands of aviation, and ensure that the necessary actions are taken now, so that the benefits can be realized later. NextGen moves away from legacy ground based technologies to a new and more dynamic satellite based technology, which will bring increased levels of traffic into our already busy airports. Smaller airports, which currently experience low levels of traffic, will feel the ripple effect; being responsible for handling the overflow of traffic that simply won't fit into the larger airports. New satellite based approach procedures will allow pilots to perform low visibility approaches into smaller airports that are typically only open during clear, unrestricted weather conditions.

> *In searching for solutions to accommodate increased "service demand" consistent with safety standards, the role of new technology is to strike a balance between the safety and efficiency of the "system" on one hand and the expenditure of public funds on the other hand.*

The 2008 spike in fuel prices, followed swiftly by the largest economic downturn since the Great Depression, has temporarily slowed the growth rate of the United States aviation industry. Nonetheless, delays continue to plague the system and will only grow worse as the number of passengers flying each year in the U.S. continues to rise. Delays resulting from the constraints of the current NAS already cost the United States approximately $9.4 billion annually, and that number will continue to spiral if nothing is done. Further adding to the stress of the current system is the emergence of new types of aircraft, including very light jets, unmanned aircraft, and commercial space transportation vehicles. New larger aircraft such as the Airbus A380 and Boeing 747-8 also create new challenges for US airports, as their larger size requires unique accommodation both while in the air and on the airport surface itself.

Many of the major focuses of NextGen are on the airborne aspects of the NAS, with regards to weather, separation, communication of aircraft. It is important that we not forget about the airport surface itself, and how the increased level of operations will dramatically affect our airports infrastructure. Particularly, how it will affect the airport itself both in terms of airport safety related issues. The Airport Safety Technology Research Program is ready to take on this challenge by introducing new innovative research programs that will greatly enhance the safety of operations at our nation's airports.

2.0 Research Direction

The Airport Safety Technology Research Program encompasses a very diverse portfolio of airport safety related programs that include visual guidance, airport capacity, airport design, surface traction, wildlife, and aircraft rescue and firefighting. Each of these program areas are very unique by themselves, but as a collective program, work together to create the safest aviation system anywhere in the world. For the next ten years, the Airport Safety Technology Research Program will have its eye set on taking airport safety to the next level by pursing several new research initiatives that will revolutionize the way airports operate today. Major areas of research include:

2.1 Winter Operations

Winter operations will become more common at smaller airports that typically had fair weather operations that only occurred during clear weather conditions. Based on the congestion expected at the larger airports, many smaller aircraft operating charter or scheduled service will be forced to operate out of smaller airports during all seasons of the year. This creates new weather and seasonal demands for operations at these airports. In addition, larger airports will have to expedite snow removal operations to reduce capacity delays as much as possible. Today, many airports struggle to keep pace with weather, requiring numerous runways to be closed for snow removal operations. This results in a ripple effect throughout the NAS as both incoming and outgoing flights affect connections and traffic delays at other airports. Maintaining operational status during snowfall events will be necessary for both small and large airports to continuously support the NextGen concept. Frozen precipitation, in the form of ice, snow, or slush, is one of the more common surface contaminations that can have a dramatic affect on the surface traction of pavement. New innovative concepts such as heated pavement, which can use either electric or geothermal heat as a heat source, show exciting potential by providing enough heat to keep the surface temperature of the runway above freezing so that any frozen precipitation melts upon contact.

2.2 Braking Performance Reporting

Braking performance during wet or reduced traction conditions will be an important factor that could affect an airport's capacity if their runway surfaces are contaminated with ice and snow. More frequent checks of the surfaces will be needed and that data could be factored into aircraft spacing landing on slippery runways to maintain capacity. Technological advances in aircraft braking systems offer optimal braking performance for modern aircraft. These systems, similar to the automated braking systems (ABS) that are equipped in most of today's automobiles, may have the ability to electronically 'share' the actions it took to decelerate the aircraft with other aircraft so that the pilot can make better decisions on whether or not he or she can land the aircraft under those given conditions. Today, we depend on visual inspection of runway surface or the use of vehicular systems that have to be physically driven on the runway surface to obtain a friction value. These systems have limited capabilities to fully assess the surface condition as they only measure the friction on a very small section of the runway. The aircraft itself could become the optimal measurement device capable of providing real time, accurate results that would be of more value to trailing pilots.

2.3 Aircraft Rescue and Firefighting (ARFF)

NextGen promotes the use of smaller airports as relievers for larger airports, and also calls for maximum utilization of those large airports that exist today. Due to the higher frequency of aircraft arrivals and departures, there is the potential for increased incidents and accidents which may lead to the need for more frequent ARFF services. There are several major considerations that need to be made. First, smaller airport fire fighting vehicles will need to be developed to enable small general aviation type airports to provide some type of fire protection service, especially with the introduction of very light jets, and increase in passenger loading on some aircraft. In addition, airports will need to be prepared to handle and respond to significantly more responses than typically expected, and will also be exposed to incidents that will likely include aircraft constructed of composite material, or aircraft carrying large amounts of cargo. While these issues are present today, we are likely to see them more prevalent over the next few years. Research must be conducted to provide a better understanding of how composite materials burn, and what can be done to extinguish the fire more effectively.

The introduction of New Large Aircraft (NLA) like the Airbus A-380 and the Boeing B-747-8, also pose new requirements for Aircraft Rescue and Firefighting services. For example, accessibility to the second level, agent requirements, and specialized equipment requirements will be higher due to increased aircraft size, increased number of operations, larger runway and taxiway complexes, and the overall increased risk of accidents/incidents due to the increased number of operations.

2.4 Visual Guidance (LED, Runway Incursions, Nanotechnology, and Holograms)

Future traffic demands will undoubtedly result in a spill over of aircraft operations onto smaller airports to relieve the larger airports from further congestion. Increased operations into smaller airports will likely require higher intensity lighting to support these operations in low visibility conditions, and may even necessitate the installation of approach lighting systems. Lower cost, longer life lighting systems that utilize LED
may be essential to support these operations at these smaller airports. Improvements in the reliability, performance, and cost of airport visual guidance devices, in regards to signage, lighting, and approach lighting, will be necessary to support the increased number of operations. New state of the art technologies such as nanotechnology and holography are just samples of the direction that visual guidance research is headed. Using very small particles that are capable on conducting electricity, nanotechnology may lead to modern advances such as paint markings and material that are capable of providing a source of electrical current. For example, it may be possible to develop a paint marking that is capable of melting snow, or an airport sign that is capable of powering itself without the use of traditional solar panels. Holography, which is a technique that allows light scattered from an object to be recorded and later reconstructed so that it appears in the same position as when it was recorded, may be utilized one day to project messages or warnings to pilots in areas where typical signage or lights are not viable. As an example, a pilot who erroneously taxi's his aircraft onto an active runway may be presented with a holographic image of a stop sign or warning message that would appear to be floating in the air out in front of the aircraft. This technology has been around for years, but has recently made significant advances in size and clarity of the images it can produce. Improving situational awareness, and reducing incidents, accidents, and runway incursions are the primary goals in visual guidance research.

2.5 Wildlife Hazard Mitigation

The Hudson River bird strike incident provided a near perfect example of how a bird strike can be catastrophic to an in-flight aircraft during take off or landing. Conservation efforts to protect the population of various species of birds, in conjunction with increased number of aircraft flying in our skies at any given time, create ample opportunities for the two to come in contact with each other. The aviation industry has made significant progress in making aircraft faster and quieter, which unfortunately makes them harder to detect by birds. Bird strikes will undoubtedly continue to be an issue for the aviation industry, and may even become more pronounced as levels of traffic into a larger number of airports continue to increase.

New technologies that can detect airborne bird activity in the vicinity of an airport are being refined to provide the timely, accurate information that is needed for airports, controllers, or even the pilot to make the necessary decisions to avoid the potential risk of a collision. New technology needs to be developed that will have the ability to disperse birds from the airport environment, and perhaps even provide a greater indication of an aircraft's presence to allow birds to see and avoid aircraft that might be approaching. Integration of systems such as these will provide valuable information to pilots to take proper action, save aircraft from being damaged by bird strikes, and greatly reduce the hazard.

Summary

NextGen is a complicated program that depends heavily on the integration of numerous systems that currently are not quite mature or ready for implementation. As the large pieces of the puzzle begin coming together to form the 'vision' suggested by NextGen, it is essential that the Airport Technology R&D Branch be prepared and ready to respond with their small piece that will fit into the larger puzzle. It is obvious that NextGen is a very advanced concept, but one thing remains very elementary: the aircraft, no matter where they fly, how fast they fly, or how close they fly to each other, will still need to land and takeoff at an airport. Therein lies our challenge.

This plan outlines the deficiencies in the current systems and procedures in each of the above areas and outlines new technologies that will be explored to assure the safety and efficiency of our air transportation system, and better enable us to meet the expected increase in air traffic. It is our hope that research outlined in this plan will lead to a demonstrable increase in airport safety.

3.0 Winter Operations on Airports: Heated Pavements

Every year, airport pavement snow removal and runway de-icing operations are an expensive component of airport winter operational costs. These operations lead to flight schedule delays that impact travel throughout the entire country resulting in postponed and cancelled flights. Chemical de-icing of pavements also leads to environmental concerns with possible contamination of nearby bodies of water. Sand/chemical mixtures also have the potential impact of creating foreign object damage (FOD) to aircraft engines. The use of heated pavements is a desirable alternative to current methods of removing snow and ice.

There are currently two primary methods for heating pavements: electrical resistance and hydronic radiant heating. Electrical resistance heating utilizes an electrical current supplied to resistive wires or conductive material embedded into the pavement to deliver heat to the pavement. The hydronic heating method involves circulating a heated fluid in tubes that have been embedded into or below the pavement. A third method for heating pavements is by applying a heat source, such as high-intensity infrared radiant heating, directly to the top surface of the pavement. Such a method is not practical at airports due to the overhead obstructions that would be necessary to apply the heat source.

By far, the greatest problem with heated pavement is that it is expensive to install and costly to operate. In an effort to reduce the installation and operating costs of a heated pavement system, various green technologies can be incorporated into the delivery of the required electricity and heated fluid. Solar panels can be utilized to provide heated fluids while solar cells can be provide electrical power. Geothermal ground loops and heat pipes can harness the heat from below the earth's surface. Advances in solar cell technology and boring practices/ equipment will help drive down the start-up costs necessary to incorporate these green enhancements into pavement heating.

Innovative concepts for heated pavements that use either electric or geothermal heat as a heat source, show exciting potential by providing enough heat to keep the surface temperature of the runway above freezing so that any frozen precipitation melts upon contact.

One of the ways that installation costs can be lowered is through the use of nanotechnology. Minute electrically conductive materials can be incorporated into asphalt and concrete mixes to provide the electrical resistance required to electrically heat pavements. It may also be possible to incorporate nanofibers into mixes to increase the heat conductivity of the pavement and to provide insulation layers to mitigate heat loss to surrounding areas.

With an expected increase in air travel by the public, greater demands will be placed upon airports to handle higher traffic loads during adverse winter weather conditions. Reducing the down time required to clear snow can be accomplished through the greater use of heated pavements. It should be our goal to begin installing heated pavements at airports frequently affected by winter weather conditions during any major pavement reconstruction project.

Specific issues that need to be addressed for heated pavements to be viable:

- Reduction in installation costs through nanotechnology.

- Reduction in operating costs through the use of green energy.

- Co-lateral use of heated pavement infrastructure with terminal operations.

- Safety of electric systems with regard to stray currents and electromagnetic interference (EMI).

- Pavement performance degradation (fatigue and thermal induced stress) from embedded conductive materials or hydronic tubes.

- Pavement reconstruction and rehabilitation procedures.

Specific Research Projects:

- Construction of an instrumented heated pavement test section at the National Airport Pavement Test Facility (NAPTF), using hydronic heating tubes and electrically resistive wires, which will be subjected to aircraft wheel loading and thermal cycling. The purpose of the study will be to measure the effect of placing such devices near the surface of an airport pavement and to develop more efficient ways of placing the pavements around the heating elements.

- A research project with a nanotechnology firm/university to investigate the use of materials to improve the electrical conductivity, thermal conductivity, and insulation properties of asphalt and concrete mixes.

- Engineering study and airport prototype installation of a hybrid heated pavement system using geothermal loops to provide heated pavement during winter weather operations and climate control (heating and air conditioning) to airport terminal buildings the remainder of the year.

4.0 Winter Operations on Airports:
Aircraft Braking Performance on Ice and Snow

Many accidents that occur during landings result from the inability of the aircraft to correctly recognize and respond to contaminated runway conditions (i.e. water, ice, slush, and snow). The ability of pilots to adequately apply aircraft braking during landings on contaminated runway surfaces is limited by the maximum ground braking friction that can be generated between the wheels and runways. Aircraft brakes are equipped with Antiskid Brake Systems (ASBS) that act to prevent wheel skidding under contaminated runway conditions with reduced ground braking friction. This project is intended to examine aircraft braking performance on contaminated runway surfaces and develop methods for more accurately predicting aircraft landing distances of follow-on aircraft on contaminated runways. It is anticipated that this methodology will allow calculation of aircraft landing distances based on aircraft performance data extracted during landings on contaminated runways. The National Transportation Safety Board (NTSB) issued a recommendation for developing the equipment and procedures to utilize aircraft performance data to predict landing distances of follow-on aircraft (NTSB Recommendations (A-07-58 through -64)[1].

The Aircraft Braking Friction Project is currently preparing the FAA's Boeing 727-25 aircraft for testing on contaminated runway surfaces. Work is being initiated to fully instrument the aircraft with strain gauges and a data acquisition system that will allow researchers to collect live data from the aircraft as it travels down and applies landing gear braking on contaminated runway surfaces. Researchers are currently developing plans for testing the aircraft on contaminated runway surfaces, and procuring equipment for an ASBS Simulation Laboratory (e.g. landing gear brake, anti-skid valve, autobrake valve, hydraulic power unit) to support development of brake system controls. When fully developed, these brake system controls will be utilized to control the ASBS on the aircraft during testing on contaminated runway surfaces. In addition, ground support equipment for the aircraft (i.e. ground

---

1. National Transportation Safety Board, Washington DC. Safety Recommendation A-07-58 through -64, http://www.ntsb.gov/doclib/recletters/2007/A07_58_64.pdf

power unit, scissor lift, and tow vehicle) has been procured. Researchers are currently utilizing the FAA Boeing 737-800 Flight Simulator, located at the FAA's Mike Monroney Aeronautical Center in Oklahoma City, Oklahoma to evaluate the simulator's ability to be utilized as a tool for simulating aircraft performance on contaminated runway surfaces.

Braking performance during wet or reduced traction conditions will be an important factor that could adversely affect an airport's capacity if their runway surfaces are contaminated with ice and snow.

It is anticipated that within the next few years, technology will be fully developed that will allow performance data collected from operating aircraft to be processed and able to accurately predict landing distances for follow-on aircraft. In order for this to be accomplished, a great deal of work will need to be done to better understand the performance characteristics of ASBS's during aircraft landing and braking on contaminated surfaces. This will require extensive testing and the development of supporting mathematical models.

Specific Project Tasks:

- Development of brake system controls in the ASBS Simulation Laboratory having operating characteristics representative of digital ASBS's, currently installed on commercial aircraft

- Integration of the developed brake system controls into the FAA Boeing 727-25C aircraft ASBS for braking friction testing on contaminated runway surfaces.

- Identification and understanding of critical ASBS parameters that influence aircraft landing performance on contaminated runway surfaces.

- Development of mathematical models that represent the performance of ASBS.

- Development of methodology for effectively communicating aircraft performance parameters to follow-on aircraft, which will allow braking performance data to be electronically shared with other aircraft landing.

5.0 Aircraft Rescue and Firefighting: Fire Fighting Operations for Cargo Aircraft

This program is a research and development effort to support needs of ARFF at Part 139 Airports that conduct cargo operations, either through dedicated cargo carriers, or through combination operations (referred to as "combi operations'). Currently, there is no reliable technically supported tactical guidance for aircraft rescue fire fighters managing cargo fires on aircraft. The National Transportation Safety Board (NTSB) has issued recommendations specifically related to ARFF preparedness for fires on cargo aircraft aircraft (A-07-100[2], A-07-101[2], & A-07-110[3]). In each of the following incidents, the use of piercing nozzles was either used effectively, attempted, or identified as a tactic that may have had beneficial outcomes if they had been deployed or deployed differently.

> FedEx 1406 – Newburgh, NY – 9/15/96
> FedEx 647 – Memphis, TN – 12/18/03
> UPS 1307 – Philadelphia, PA – 2/17/06
> FedEx 630 – Memphis, TN – 7/30/06
> ABX Air 767-200 – San Francisco, CA – 6/29/08

This project includes full scale fire testing in various Unit Load Device (ULD) types and configurations in aircraft main deck and lower deck holds. Significant research is needed to support theories regarding High Reach Extendable Turret (HRET) tactics and strategies. Piercing tip length standards and guidance do not exist for cargo fires. The distances from the outside fuselage skin to the outside walls of cargo ULD's range from 12" to 57". Piercing tip lengths range from 22" – 34". This research will measure the effectiveness of agent application methods outside of ULD, effects of different agents including clean agents.

Airports will need to be prepared to handle and respond to significantly more responses than typically expected, and will also be exposed to incidents that will likely include aircraft constructed of composite material, or aircraft carrying large amounts of cargo.

Over the next ten years, this research project will provide the technical data to support the operational guidance provided to the ARFF at Part 139 Airports that conduct cargo operations. The numerous variations in which the ULD's are configured within the cargo aircraft create a challenge to responding ARFF units. Current federal regulations vary between items that can be shipped on passenger aircraft and cargo aircraft. There is a large misunderstanding in how to combat fires within cargo aircraft for items such as Lithium Ion battery packs which airport fire fighters would not see on passenger carrying aircraft.

---

2. National Transportation Safety Board, Washington DC. Safety Recommendation A-07-97 through -103, http://www.ntsb.gov/doclib/recletters/2007/A07_97_103.pdf
3. National Transportation Safety Board, Washington DC. Safety Recommendation A-07-110, http://www.ntsb.gov/doclib/recletters/2007/A07_110.pdf

Specific Project Tasks:

- Determine the effects that the interior lining material currently used to line the fuselage of cargo aircraft may have on both direct and indirect fire attacks. Research will be conducted to better understand lining materials burn characteristics, interaction with fire fighting agents, cooling properties, and resistance to penetration or cutting tools.

- Determine the effects that multiple ULD loading configurations can have on interior fire attack. Research will be conducted to document the various possible ULD configurations being used on today's commercial and cargo aircraft, and to identify key strategies that fire fighters can use to better identify the layout of ULDs that they can expect for a given aircraft model, and better apply agent to extinguish the fire as rapidly as possible.

- Evaluate the effectiveness of aircraft skin penetrating nozzles on fighting fires in the interior of cargo aircraft. Research will be conducted to determine the optimal nozzle length, spray pattern, entry locations and techniques for operation of a skin penetrating nozzle for fighting fires located inside the fuselage of a loaded cargo aircraft.

- Develop guidance on fire fighting tactics and agent requirements for combating Hazardous Cargo fires. Research will be conducted to develop guidance material for airport fire fighters that will provide detailed information on the best techniques for applying agent, the best attack method, safety issues and precautions, and the importance of using personal protection equipment (PPE).

6.0 Next Generation Aircraft Rescue and Firefighting:

6.1 Advanced Composite Materials Firefighting

Over the past decade, the use of composites for aircraft structures has dramatically increased. Both Boeing and Airbus are relying more heavily on the use of composites to lower the weight of the next generation aircraft. Boeing is the first aircraft manufacturer to design a fuselage made completely from composites. The Boeing 787 Dreamliner is more than 50% composites by weight. In comparison, the Airbus A380 has a variety of composites totaling 22% by weight, with plans to expand the use of composites up to 40% in later versions of the A380. Research related to fires involving composite materials to date have mainly focused on the health hazards associated with particulate and toxic gas generation. However, limited information is available on the effectiveness of fire fighting agents and techniques used to extinguish composite fires, especially on NLA.

Under the current research program, testing has primarily focused on the development of a live-fire test protocol. Over the next ten years, research will continue to further develop and refine this protocol into an industry accepted standard. This protocol can then be utilized to evaluate the various fire agent types, quantities and delivery technologies that may be necessary for assisting our nation's ARFF departments in combating these types of fires.

6.2 Biofuels

Recent developments in biofuels, which are fuels whose energy is derived from biological carbon fixation, are likely to have a significant impact on aviation over the next 10 years. Reduced emissions, lower costs, and less dependency on crude oil are very attractive benefits that our nation's airlines are likely to take advantage of. As a result, research must be conducted to ensure that our current inventory of extinguishing agents is capable of suppressing fires involving biofuels. Agent effectiveness, extinguishment times, and interaction with biofuel are just a few of the many areas that will be researched.

Specific Research Projects:

- Identification of extinguishing agents for use on composite material fires. Extensive research will be conducted to identify firefighting agents that possess the best capability for attacking, extinguishing, and suppressing fires involving composite materials. Characteristics such as cooling ability, quantity required, and ability to contain composite micro-fibers will be fully investigated.

- Determine the effectiveness of commercially available forcible entry tools on advanced composite materials. Research will be conducted to evaluate several different types of hand held forcible entry tools that are currently available commercially to determine which types of tools possess the quickest penetration rate, ease of use, and overall performance in various types of composite material being used to construct aircraft of today and tomorrow.

- Determine the effectiveness of commercially available rescues saws and various blade types on cutting composite materials. Research will be conducted to compare the performance of various makes and models of rescue cutting saws, along with various cutting blades, to identify the optimal combinations that will offer airport firefighters the quickest, safest, tool combination for gaining entry into an aircraft constructed of composite materials.

- Researching the potential effectiveness of boom mounted cutting tools. Research will be conducted to investigate the concept of utilizing cutting tools mounted to a fixed high reach extendible turret to remotely cut through the fuselage of an aircraft. The feasibility of using such a device will be explored through the development of a prototype unit and full scale testing.

- Extensive research will be conducted to the best technique for fire fighters to use for attacking, extinguishing, and suppressing fires involving composite materials. Research will include determination of the best application rate, placement of agent, suppression techniques and containment of composite micro-fibers.

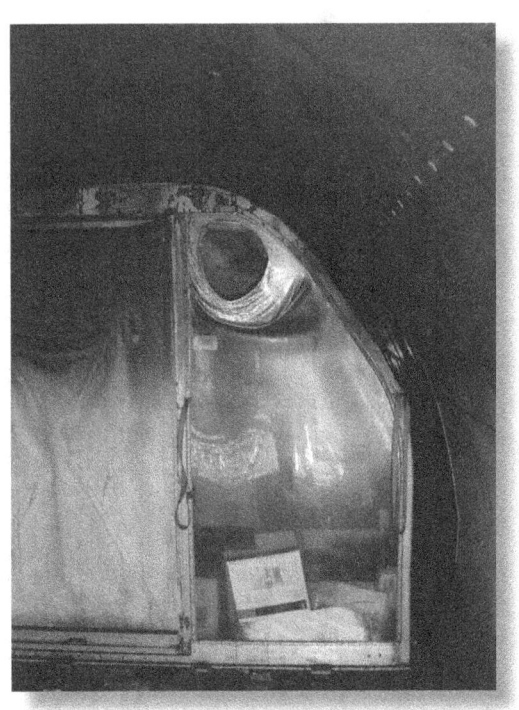

- Determine baseline data for composite fires to provide researchers with comparative values for future composite firefighting research. Through the use of full scale fire testing, research will be conducted to obtain data the time and amount of agent that is required to completely extinguish composite fires of a given size. With this data, researchers will better understand the characteristics of a composite fire so that new agents or application techniques can be compared to baseline data.

- Identifying the numerous hazards associated with the toxic fumes and airborne composite fibers on the respiratory systems of both evacuees of an aircraft and the ARFF personnel responding to the incident. Extensive research will be conducted to provide full documentation of the hazards of composite materials, and what can be done to mitigate the risk of contamination through use of breathing apparatus, PPE, filtering mechanisms, application of adhesive fluids that contain the fibers, etc.

- Determine if existing extinguishing agents are capable of suppressing fires involving biofuels. Agent effectiveness, extinguishment times, interaction with biofuel effects on ARFF equipment and personnel, etc. will also be investigated.

7.0 Aircraft Rescue and Firefighting: Second Level Access for NLA

The Airbus A-380 and the Boeing 747-8 present unique challenges for the airport firefighting departments responsible for protecting their respective airports. This research effort focuses on specialized vehicles and/or equipment needed to provide firefighters with the proper tools needed to attack, access, extinguish and rescue passengers from an NLA. Today's airport firefighters respond to different types of emergencies involving aircraft such as exterior fires, interior fires, rescue, passenger deplaning, medical response, security, etc. Due to the variety of ARFF responses, there has been an increasing trend of airport fire departments purchasing air stair vehicles or other specialty vehicles to access an aircraft. The FAA has been actively researching a new concept vehicle called an Interior Access Vehicle (IAV) for rapid access to aircraft doorways to aid fire fighters in making a safe and rapid entry into an aircraft fuselage, as well as assist in the egress of passengers.

---

*Accessibility to the second level, agent requirements, and specialized equipment requirements will be higher due to increased aircraft size, increased number of operations, larger runway and taxiway complexes, and the overall increased risk of accidents/incidents
due to the increased number of operations.*

Much of the research to date has been conducted with a significant amount of industry involvement. Studies have been conducted on how airport fire departments have addressed the needs for these vehicles, concepts of vehicles modified to accomplish these tasks and events around the world where these vehicles were or could have been utilized. Early design standards and requirements have been created by the FAA with the assistance of the National Fire Protection Association ARFF Technical Committee. Now that standards exist, the vehicle manufacturers are in the design stages of vehicle development. Over the next 10 years, research will continue on the design and development of IAVs.

Specific Research Projects:

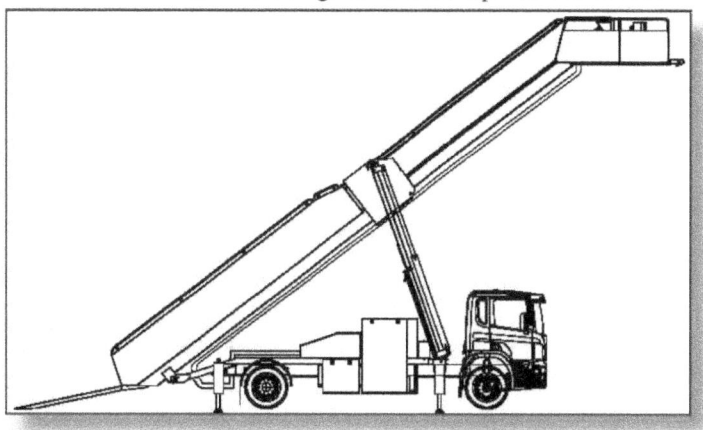

- Research historical data collected on current and past aircraft rescue events around the world where interior intervention type vehicles were or could have been utilized. This data will be utilized to assist in the refinement of new vehicle design concepts, and potentially for future development of standards.

- Establish complete design concept for a IAV, through cooperation with leading rescue vehicle manufacturers. Research will be conducted to develop prototype IAV from concept through prototype development, and conduct necessary performance testing of prototype IAV vehicle to develop performance standards for these types of vehicles.

8.0 Visual Guidance: Safety, Capacity, and Environment

With increasing fuel prices and the potential of increased delays at airports as the number of passengers flying continues to rise, it is imperative that all users of an airport (pilots in aircraft and drivers in ground vehicles) know where they are, know where they are going and know how to get there safely and efficiently. Delays cost the United States approximately $9.4 billion annually.

With the development of new and improved airport signage which provides location information (know where you are), markings and lighting (know where you are going and how to get there) for direction, improved situational awareness can help decrease runway incursions, decrease delays and increase capacity.

Improvements in the development of nanotechnologies, the concept of having airport signs able to generate their own power for illuminating themselves may be attainable. This technology also promotes the possibility of providing addressable signage for specific route guidance for aircraft.

New state of the art technologies such as nanotechnology and holography are just samples of the direction that visual guidance research is headed. Using very small particles that are capable on conducting electricity, nanotechnology may lead to modern advances such as paint markings and material that are capable of providing a source of electrical current.

Advancements in holographic technologies may be used to enhance markings on airports to aid in the elimination of runway incursions. Using holograms could create a 3 dimensional marking which would be more conspicuous to the user. The advancements in this technology warrant research in this area.

With the development of reliable logic to control lights as in the Runway Status Light System (RWSL), and new light sources, the ability to provide lighting for each individual aircraft can be attained. The concept of controlling taxiway lighting was developed in the late 1990's; however, adequate technology did not exist to implement this Advanced Taxiway Guidance System. This system would illuminate the taxiway lighting in front of an aircraft, and extinguish the lights behind the aircraft automatically. This would help alleviate the "sea of blue" effect of all taxiway lights being illuminated, that pilots have mentioned when trying to determine their taxi route thus providing visual cues specific to that pilot.

To forward this capability, research is being conducted to develop a Low Cost Surface Surveillance Framework (LCSSF). This framework, just like NextGen is developing enroute navigation not based on radar, will identify a system for the purpose of surface surveillance using a variety of sensors to provide very reliable surveillance without the need for costly radars.

Incandescent lighting has been used in aviation for over 60 years. This lighting technology has provided the needed cues to provide a safe environment to conduct aviation activities. This technology though, is very inefficient in producing light in terms of amount of light (lumens) versus energy required (watts). The incandescent source produces mostly wasted energy in the form of heat compared to the amount of light produced. Until recently there has not been another alternative. Recent developments in Light Emitting Diode (LED) technology has enable their use with the advantage of producing mostly light (lumens) with very little wasted energy (watts). This technology also provides a longer life time between failures of approximately 30,000 hours compared to the incandescent source which is approximately 2,000 hours.

Specific Research Projects:

- Development of new specifications for Heliport lighting that will include new edge and inpavement lighting configurations for optimal landing pad identification, as well as detailed photometric and color standards for the optimal lighting fixtures to be used.

- Development of new visual cues to provide sufficient visual conspicuity for pilots conducting GPS approaches into Heliports. Research will be conducted to identify the optimal configuration, intensity, color, and spacing of the lights to be used to provide the visual portion of the GPS approach.

- Research will be conducted to identify the specifics required to support the FAA's Low Cost Surface Surveillance Framework (LCSSF), which will provide the necessary foundation from which vendors can develop lower cost surveillance systems capable of providing data similar to that currently produced by more expensive ASDE X systems.

- Research will be conducted to further develop an Advanced Taxiway Guidance System utilizing LEDs lighting and RWSL logic. Evaluation of various sensing technologies, as well as lighting control sytems, will be conducted to identify the optimal configuration necessary to turn lights on and off to direct aircraft around the surface of the airport.

- Development of a "smart sign" that utilizes nanotechnology to produce its own power so that the sign is self sufficient, without reliance on energy from an airport lighting circuit. Additional applications for nanotechnology will also be explored.

- Research will be conducted to develop holographic equipment capable of generating images of enhanced airport signs or warning message on the airfield.

- Research will continue to develop improvements in surveillance and visual cues for use in the airport environment. Evaluation of new lighting technology, including next generation light emitting diodes, solar energy, lower voltage lighting, higher visibility marking material, and addressable message signs will be initiated as the technology becomes available.

9.0 Wildlife Hazard Mitigation

Collisions between aircraft and wildlife have occurred since the beginning of manned flight just over a century ago. The last 50 years have seen an increase in both number of occurrences and the severity of damage resulting from strikes. The reasons for the increases include the introduction of faster, quieter aircraft, increased airspace capacity, and significant increases in bird populations. During this time, the FAA's wildlife hazard management program has focused on mitigating wildlife hazards on or near airports through various methods including habitat modification, harassment technology, partnerships with academia, military, government, and the aviation industry.

Research is currently focused on reducing the risk of collisions between aircraft and wildlife by identifying, studying, testing and developing methods, tools and techniques in four main areas:

- Habitat Management – airport property can be attractive to hazardous species of wildlife for a variety of reasons including vegetation and open bodies of water. Some smaller general aviation airports in rural locations offset costs by leasing land for agricultural use. Research in this area is aimed at identifying ways in which necessary land use on airports can be managed in a manner that makes it least attractive to hazardous species.

- Control Techniques – despite best efforts to keep hazardous species off of airport property, wildlife can still be present. Current research in this area is covers methods such as fencing, that limit access to safety areas, and tools for deterring and harassing wildlife in order to influence their dispersal from the area.

- Wildlife Strike Analysis – The FAA currently maintains a database of report wildlife strikes. Data collection and analysis continues to yield information about the circumstances surrounding strike events. With this knowledge, appropriate mitigation efforts can be applied. This component of the wildlife research program includes collaboration with the Smithsonian Institution's Feather ID lab for identification of bird strike remains for species determinations.

- Bird Detection and Deterrence Technology – Assessments of commercially available bird detection radar systems have lead to the release of an Advisory Circular on Avian radar use on civil airports. The current systems however have limitations while other military based systems continue to emerge.

The next 10 years will see a gradual completion of some research projects with continuation and adaptation of many others toward emerging problems. The outlook is for relative reductions in the efforts dealing with habitat management and significant increases in the evaluation, testing, and development of technology solutions to detect and deter wildlife. Efforts will continue toward filling gaps in bird strike reporting including implementation of push/pull integration services between discrete databases.

Specific Research Projects:

- Research will be conducted to assess the hazard level of biofuel and forage crops being grown on and near Airports. Attraction to birds, rodents, and other wildlife will be evaluated and assessed to see if these types of crops attract or deter wildlife.

- Research will be conducted to assess the of hazard level of solar farms on and near Airports. The construction of large scale solar farms on airport property potentially creates an attractive habitat for birds and other rodents, as the large solar panels provide ample protection from predators, as well as numerous roosting and nesting structures. Evaluation of wildlife activity will be conducted to determine if a hazard exists.

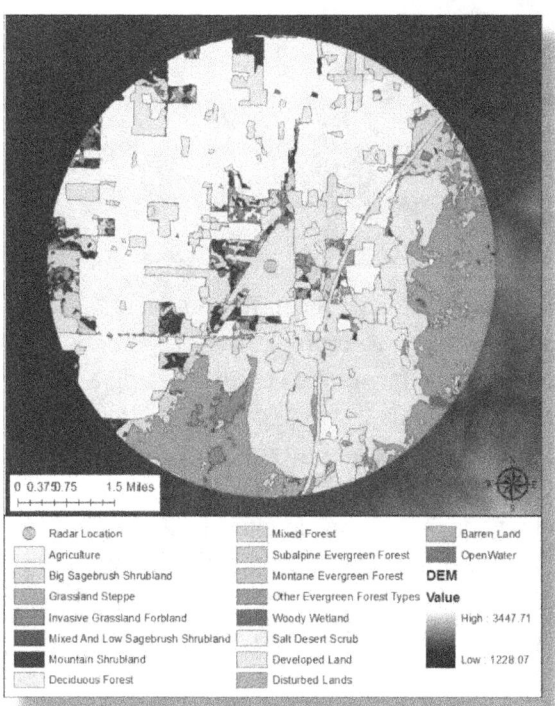

- Research will be conducted to explore the potential use of avian radar systems as a tool for conducting wildlife hazard assessments. These ground based systems offer the potential for being used as a tool to conducting assessments of potential hazards by identifying and tracking bird activity around the radar site.

- Assessments of Radar Based Detection Systems Augmented with Directed Energy Deterrent Mechanisms will be conducted to explore their feasibility in dispersing birds from the immediate airport area, potentially eliminating the risk of a collision of a bird with an aircraft. These systems will be thoroughly evaluated to determine their effectiveness, accuracy, safety of use, and cost effectiveness.

- Development of Holographic Barriers Perceivable to Avian Species. Research will be conducted to develop holographic equipment capable of generating images on an airport surface that could be used to deter birds from landing or foraging on the airport.

- Research will continue to refine the acquisition of bird strike data from discrete databases that contain ample amounts of information on bird strike activity as it is reported from pilots, airlines, airports, or wildlife biologists. As data continues to populate these databases, research must be done to provide database users with easy to use, real time data figures that can be used to take wildlife mitigation actions

- Research will be conducted to integrate avian radar information into air traffic control procedures so that controllers can provide real time information to pilots approaching or departing from an airport. With the live, avian radar data, controllers may be able to vector aircraft around areas detected to have heavy bird activity, providing the necessary separation required to avoid in flight collisions

- Operational analysis of deployed avian radars will be continue, allowing researchers to better understand how various detection technologies perform under various weather and operational conditions. Data collected under this research effort will be used as required to modify avian radar performance standards.

- Research will be conducted on the potential integration and fusion of local, regional and national radar data for wildlife hazard forecasting and real time activity information. Development of algorithms that will merge data collected from these various radar sources could provide a single stream of data capable of providing users with complete national coverage of avian activity that can be used in flight planning, forecasting, and better understanding of migratory behavior.

- Assessments of military target detection systems and weapons systems adapted for wildlife hazard application will be conducted to determine if they can be utilized in the civilian world to detect and deter birds. These types of systems will be evaluated to see if they offer improved resolution, faster or higher accuracy detection, or a lower in cost that those system being developed specifically for avian detection.

10.0 Airport Planning & Design:

10.1 Airport Planning

The Airport and Airspace Simulation Model (SIMMOD) simulates aircraft operations both in the airspace around an airport and on its runways, taxiways, and gate areas. This helps the Federal Aviation Administration and an airport operator evaluate which airfield, terminal, or aircraft operational improvements would provide the greatest capacity enhancements. The model also helps airport users, such as airlines and cargo carriers, gain greater efficiency in their services to the aviation public. The FAA and the aviation community use the model extensively worldwide to reduce delays, increase airport capacity, and improve the overall efficiency of the National Airspace System (NAS).

Research is currently focused on airport & airspace simulation enhancement, building the airport database, improving digitization of airports, and developing new utility programs. Details of the four main areas are as follows:

1. Airport & Airspace Simulation Enhancement – consists of analysis of airspace interactions between arrivals and departures for a given airport, analysis of airspace interaction between multiple airports, and computing aircraft travel times and delays (to compare and improve airport or airspace network with the baseline).

2. Building Airport Database – consists of updating the database with new runways, taxiways, gates, etc. for digitizing of airports and physical onsite data collection at airports.

3. Improve Digitization of Airports – consists of making revisions to existing suite of fast-time models (including the FAA's Airport and Airspace Simulation Model) to accept the new digitized setup and developing a GUI front to existing fast-time simulation models.

4. Develop new utility programs – Consists of end-testing utility programs to perform the following functions: Modifying specific flights of an existing schedule, Supplementing traffic schedule with additional flights to simulate future demand levels, and converting traffic schedule to ADSIM/ RDSIM format.

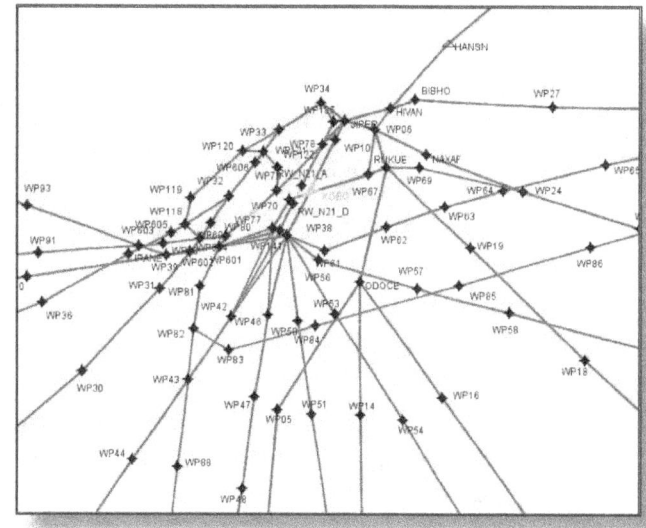

Over the next 10 years, extensive work will continue to support and refine the FAA's airport planning and capacity programs to support the traffic demands forecasted in the NextGen program. The goal will be update capacity models in a pace that will match the gradual implementation of new approach and departure procedures, utilization of small airports, and closer separation standards.

Specific Research Projects:

- Software updates for the airport and airspace simulation enhancements engine will continue to be updated and released every March and September of each year.

- Efforts will be made to generate an automatic reporting capability for outputs from the simulation engine.

- Efforts will be made to improve the visualization on integrated platforms within the simulation software.

- Research will continue to expand the usability of Official Airline Guide (OAG) data, as well as integration of several other Collaborative Decision Making (CDM) tools such as the Enroute Radar Intelligence Tool (ERIT), Enhanced Traffic Management System (ETMS), the National Traffic Management Log (NTML), and the PASSUR predictive analytics software program.

- Programming efforts will continue to merge ERIT and ETMS data to collectively decipher beacon codes from aircraft.

- Develop capability to use SIMMOD and GIS applications to enhance digitization of airport data.

- Develop capability to display playback animation with geographical software applications such as Google Earth or Visual Earth to allow users to obtain 2-D airport animation from top view, 3-D airport animation from tower, and 3-D metro airspace from different angles.

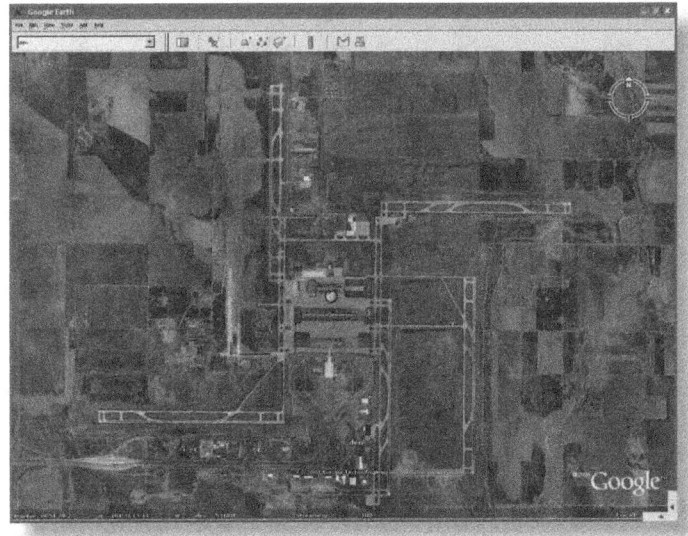

- Research will continue with activities such as assigning special equipage to specific airborne flights to perform assessments of new, emerging technologies as they are developed.

- Efforts to update standard aircraft profiles with more current takeoff weight parameters so that actual fuel consumption can be better understood and integrated and more accurately depicted in modeling software.

10.2 Aircraft Noise and Sleep Annoyance Research

Aircraft noise is generated through the various stages of flight, both on the ground and in the air. The impacts of aircraft noise are used for tasks such as determining land-use compatibility guidelines around airports and noise mitigation funding. Early research conducted by T.J. Schultz[4] (4) in 1978 is the basis for all current aviation noise policies in the United States. Schultz developed a correlation between transportation noise exposure levels in terms of the day-night average noise level (DNL) and the percent of the population highly annoyed. For his study, Schultz used existing social surveys on noise annoyance conducted in the 1960's and 1970's from a variety of countries. Schultz also suggested that annoyance from the total effect of many quiet events is equal to the annoyance from a few louder events. In 1992, the Federal Interagency Commission on Noise (FICON) re-affirmed Schultz's work, yet stated, "This work is continuing and may provide a basis for an improved understanding of community response to noise." Not only is the Schultz data 30 years old, but the research also included multi-modal transportation (air, rail and road) and was conducted at a time when aircraft operations were louder and less frequent.

Current data available shows that people react more adversely to aircraft noise than to noise from other modes of transportation. In addition, noise concerns do not appear to have abated in spite of the substantial reduction in DNL 65 contours nationwide. The latest research specific to the aircraft noise dose-response relationship has been done in Europe and Asia; current U.S. data has only been obtained in response to lawsuits against airports, which may not be a true reflection of people's annoyance levels. Since noise is the most immediately objectionable community impact of aviation and requires the most Federal resources to mitigate, it is crucial to collect updated community annoyance data. By conducting a broad survey to update the Schultz dose-response curve for U.S. airports, one could gain additional confidence that the models are correct for modern aviation noise in today's setting.

Over the next 10 years, research will be conducted to update the scientific evidence of the relationship between aircraft noise exposure and its effects on communities around U.S. airports. This will require conducting social surveys and collecting noise data to measure subjective reactions to aircraft noise and characterize community noise exposure across a broad spectrum of airports with a wide range of aircraft noise exposure and responses.

Specific Research Projects:

The following list contains the major areas of focus:

- Research will be conducted to develop a survey instrument that will be established to obtain dose-response data from the public to determine their annoyance level.

- A list of candidate airports will be established to participate in the study, reflecting a variety of geographical locations and sizes. Involvement in noise issues and/or availability of noise monitoring data will assessed, as well as recency of noise measurements, number of runways contoured, amount of historical noise data available, and mix of aircraft.

---

4. Schultz, T.J. (1978). "Synthesis of social survey on noise annoyance," J. Acoust. Soc. Am 54, 377-405

- Human response data and noise exposure data within communities exposed to aviation noise will be conducted at several airports, using an approved survey instrument. The surveys shall be conducted out to DNL 65 noise contour in both rural and urban areas surrounding each selected airport. Surveys out to the DNL 55 noise contour will also be conducted. When possible, noise exposure data will be collected from existing noise monitors already installed at selected airports, newly established noise monitors, and/or a combination of noise monitors and environmental software modeling.

- A final report will be developed after analysis of noise and response data is conducted, to create new dose-response curves for annoyance.

11.0 Airport Technology Research Taxiway

Our nation's airport lighting, markings and signs infrastructure has become very old, operates very inefficiently, and is based on operational data and technologies that are now decades old. A significant amount of electrical energy is needed to power these systems costing airport operators significantly more money. Compounded with the aging equipment, wiring, and connections, the cost to maintain the systems becomes very significant. Under this project, the FAA would undertake the task of developing a state of the art research test bed that would provide Airport Safety Technology R&D Sub-team engineers an opportunity to design, install, test, monitor, and report on the construction, operation and maintenance of research infrastructure that takes full advantage of state of the art technologies in signs, lighting and markings.

Major advances in airport technology have brought forth new brighter, more efficient and more conspicuous lighting devices, enhanced paint material that lasts longer than traditional paint, and airport signage that is easier to read from greater distances. This new technology, when compared with the current state of legacy systems, warrants that the FAA undertake a major research effort to enhance these essential systems, making improvements that will best serve the future of our nations aviation. The FAA's "NextGen" Program talks about levels of air traffic increasing to three times what it is today, bringing thousands and thousands of aircraft to smaller airports that have historically seen very little traffic. The demand for the infrastructure at these airports will increase significantly, bringing with it higher levels of usage, higher performance requirements, and higher costs to maintain. Today's General Aviation community is already indicating that there is a need to enhance their visual aids, citing examples of aging power cables, antiquated fixtures, and high energy costs as major problems that they are experiencing now. This problem even extends out to our remotely located airports that are far away from commercial power sources. In many cases, these airports are open only during daylight hours, or only allow operations when sufficient notice is given to power a diesel generator to light the landing area for an approaching aircraft.

At the present time, Airport Safety Technology R&D Sub-team engineers depend heavily on volunteer airports that are willing to allow the FAA to temporarily test a new device or material. This is usually done on very small scale, where perhaps a few lighting fixtures, or a few paint stripes are placed in an obscure area of the airport where their placement would not dramatically affect aircraft operation. There are usually a series of waivers, or special approvals that are required to allow these types of evaluations. Sometimes, these requests can take well over 9 months to obtain approval for each separate test site. In many cases, the airports that volunteer their facilities for evaluations like this are located many miles away from the FAA's Technical Center, and as a result, require significant amount of time and resources for each site to be accessed. Research activities are limited to occasional visits or reliance on reports from local airport personnel.

Under this project, we will develop a single test site that would allow Airport Safety Technology R&D Sub-team engineers to design, install, test, monitor and report on the performance of airport technology in a single test area on a continual basis. One consideration was to find a moderately sized airport that maybe be a reliever facility for NextGen with a runway, approximately 5,000 to 7,000 feet in length, that could have its entire infrastructure overhauled, and reconstructed with the more technologically advanced visual aids possible. This could include lower voltage wiring, LED lighting fixtures, new generation paint material, essentially bringing the best technology available into one location to create this infrastructure. Another consideration was to construct the test area such that it could be powered solely by a renewable energy source, making it the most environmentally friendly airport in the nation. Researchers would have the opportunity to explore solar or wind energy as viable power sources that would have the capability of supplying sufficient energy to power the airport infrastructure without any reliance on commercial power, with no pollution, in a completely self-sustaining system. Leading edge technologies, such as holograms, photo luminescent paint and nanotechnology are two developments that may have great applications in the airport environment. Imagine painting a runway centerline with paint that glows all night long without any lighting or power, or imagine that an airport sign could be painted with material that itself generates power, making the sign structure itself its own power supply. Again, research engineers need a test site where this technology can be put into use for research and evaluation purposes, without the need to travel to many different test sites willing to assist in evaluating components of this infrastructure.

To that effect the FAA Airport Technology R&D Team has entered into a Memorandum of Agreement (MOA) with the Delaware River and Bay Authority (DRBA). The MOA grants the FAA the right to construct, and operate and maintain research infrastructure at Cape May County Airport (WWD) in Erma, New Jersey.

Specific Research Projects:

Major challenges that will need to be addressed to reach the requirements of this project include:

- Rehabilitation of a decommissioned taxiway in order to develop a state of the art research test bed that will allow for airport safety and pavement research initiatives

- Design to be consistent with existing airport conditions and be utilized as a standard taxiway by the airport when the FAA is not actively conducting research on the test bed.

- Necessary data acquisition systems, monitoring equipment, etc. would be procured and installed to complete the test site, giving researchers sufficient data tools to operate and monitor the various technologies that were installed.

- During the construction of test pavements, extensive in-situ testing (including non-destructive tests) will be performed to characterize material properties of as-constructed pavements.

Airport Pavement Technology Research and Development

1.0 Introduction

Figure 1: Cover of first R&D Plan brochure.

In April of 1993, the Airport Technology R&D Team published a research and development plan to improve pavement design and evaluation titled *"Airport Pavements – Solutions for Tomorrow's Aircraft,"* (reference 1). The plan addressed what was then seen as a crisis in credibility of the existing airport pavement thickness design procedures to predict the damage that would be caused to pavements by the next generation of aircraft. It was suspected that the design procedures were overly conservative. But there was no rational means of determining whether this was in fact true, and if it was, to what degree. The aircraft which called into question the credibility of the design procedures was the Boeing B-777, with its very heavily loaded six-wheel landing gears. In the case of flexible pavements, the existing design procedure incorporated a purely empirical extension of the basic mechanistic response model which was dependent on the number of wheels in a gear. The empirical factor for six-wheel gears had never been explicitly derived from full-scale traffic tests and confidence was low that new pavements designed with the procedure would have the predicted life. Consequently, plans for accommodating the B-777 in the existing fleet were based on the most conservative assumptions. This led to predictions that approximately one half of the existing flexible pavements would need to be strengthened at very high cost and that new flexible pavements would need to be significantly thicker than was needed for the existing fleet. In the case of rigid pavements, the response model for the existing design procedure was based on computing edge stress of a semi-infinite slab, whereas the B-777 six-wheel gear extends, lengthwise, over 50 percent of a typical slab. And the total load on a single slab from a six-wheel gear would be 50 percent higher than from a four-wheel gear with the same wheel load. As for flexible pavements, full-scale traffic test data was not available to determine the effect on life of this difference between the design assumptions and field conditions. The plan (reference 1) therefore concentrated on improving the representation of the pavement structures in the design response models and collecting full-scale traffic test data with loads applied to represent the new six-wheel gear configuration.

> *...total funding required to execute the plan was $55 Million which included costs for construction and operation of the necessary full-scale testing equipment; the plan was the FAA's largest and most significant pavement R&D effort.*

Execution of the plan was coordinated with the FAA Office of Airport Safety and Standards (AAS), with AAS being responsible for the preparation and release of advisory circulars containing products of the work done under the plan. The contributions of AAS personnel are gratefully acknowledged, particularly those of Harold Smetana, James R. White, John Rice, Jeff Rapol, and Rodney Joel. An industry working group, co-chaired by Ed Gervais, also provided help and guidance throughout the period. And, as always, thanks to Dick Ahlvin for his many contributions.

Since publication of the plan in 1993, the following milestones have been achieved:

1. Publish a design and cost study for a full-scale trafficking test facility, DOT/FAA/CT-93/51, "Airport Pavement Test Machine Design and Cost Study" was published in October, 1993.

2. Develop a Microsoft Windows-based computer program with a layered elastic response model. The program was called LEDFAA and was published concurrently with a new advisory circular, AC 150/5320-16, "Airport Pavement Design for the Boeing 777 Airplane." The AC was published on October 22, 1995, and required the use of LEDFAA for flexible and rigid pavement design of any pavement intended to serve the B-777.

3. Design, build, and put into operation a full-scale facility for traffic testing airport pavements subject to loading by landing gear configurations representative of the latest generation of aircraft. The National Airport Pavement Test Facility (NAPTF), located at the William J. Hughes Technical Center, was commissioned and dedicated on April 12, 1999. The test

Figure 2: NAPTF test vehicle in the facility.

facility design and construction was a joint venture with the Boeing Company under a Cooperative Research and Development Agreement (CRDA). Two thirds of the cost was funded by the Government and the remaining third funded by the Boeing Company. The CRDA also provided for collaboration on test planning for a period of fifteen years.

4. Develop revised traffic failure models for LEDFAA based on flexible and rigid pavement test data obtained from the NAPTF. LEDFAA version 1.3 was published in conjunction with Change 3 to AC 150/5320-6D, "Airport Pavement Design and Evaluation," and the cancellation of AC-150/5320-16. Change 3 to AC 150/5320-6D fully incorporated LEDFAA and allowed its use as an alternative for design of any airport pavement covered by the AC. The use of LEDFAA for pavements serving B-777 aircraft was still essentially mandatory because design charts for the B-777 by the old design procedures had not been included in the AC.

5. Develop new "alpha factors" for the computation of pavement classification numbers (PCN). ICAO published new alpha factors for the computation of PCN of flexible pavements on October 16, 2007. The new alpha factors are based on a new set of full-scale test data incorporating test results from the NAPTF. Descriptions of the test results and derivation of the new alpha values are given in a series of FAA reports: DOT/FAA/AR-06-7, "Alpha Factor Determination Using Data Collected at the National Airport Pavement Test Facility," March, 2006; and DOT/FAA/AR-08/01, "New Alpha Factor Determination as a Function of Number of Wheels and Number of Coverages," January, 2008.

6. Develop a new response model for rigid pavement design based on 3D finite element techniques. A beta version of a new computer program, called FAARFIELD, intended to replace LEDFAA, was published in June of 2007. Rigid pavement design in the program replaces the layered elastic response model of LEDFAA with a 3D FEM model. The rigid pavement failure model was modified from the LEDFAA model to match the different stress response characteristics of the FAARFIELD FEM response model. The new failure model was based on an analysis of the existing set of historical and NAPTF full-scale test data. The flexible pavement response model is unchanged from LEDFAA. Subsequently, on September 30, 2009, AC 150/5320-6D was cancelled and replaced by AC 150/5320-6E. The new AC completely eliminates reference to the old, chart-based, design procedures and requires the use of FAARFIELD in place of LEDFAA for the design of both flexible and rigid pavements.

7. Develop an upgraded and enhanced pavement performance and monitoring system (PPMS). A web-based PPMS computer program has been developed. The program, called FAA PAVEAIR, was released in February 2011 in beta version.

Following these milestones, the FAA has completely eliminated the use of the old design procedures and replaced them with modern, computer-based, models fully validated for the new generation of aircraft with full-scale test data. The new generation has expanded from the 537,000 lb loading of Boeing's B-777 of 1995 to include:

1. 777,000 lb B-777-300 ER, two six-wheel gears, 59,400 lbs per wheel.
2. 1,255,000 lb Airbus A-380, two six-wheel and two four-wheel gears, closely spaced, 59,600 lbs per wheel.
3. 973,000 lb B-747-8, four four-wheel gears, closely spaced, 57,800 lbs per wheel.
4. 840,400 lb A-340-600, three four-wheel gears, 66,500 lbs per wheel.

Although the full 20-year design life of new pavements built to serve these aircraft has not passed since their construction, there have been no reports of premature pavement distress or failure associated with operation of the aircraft. Therefore, with the exception of some future improvements to be discussed later, the latest airport pavement design and evaluation advisory circular, AC 150/5320-6E, is considered to be suitable for thickness design for the current fleet of heavy aircraft with a high degree of confidence in the thickness computations as affected by gear configurations and wheel loads.

> *...the latest Advisory Circular, AC 150/5320-6E, is considered the most up-to-date document providing the highest degree of confidence in the airport pavement thickness design computations.*

Other aspects of pavement design, construction, and evaluation, such as mix design, material selection, environmental factors, maintenance, sustainability, and life-cycle description, need to be studied and procedures developed in the light of changing governmental responsibilities and fiscal regimes. In addition, the AAS business plan for 2011 included an item to evaluate the feasibility of extending the design life of runway pavements at major hub airports from 20 years to 40 years, with, presumably, the expectation that life extension would be required for other airport pavements if it is shown that the concept is feasible and practically achievable. This report lays out a 10-year plan for studying these aspects of airport pavement operations, starting with a discussion of how to define pavement life under these new conditions and why a more precise definition of life-cycle cost analysis is required to satisfactorily meet evolving requirements over extended periods of time.

2.0 Airport Pavement Life

Advisory circular AC 150/5320-6E states in paragraph 303.a. that:

"Pavements designed and constructed in accordance with FAA standards are intended to provide a minimum structural life of 20 years that is free of major maintenance if no major changes in forecast traffic are encountered. Rehabilitation of surface grades and renewal of skid-resistant properties may be needed before 20 years because of destructive climatic effects and the deteriorating effects of normal usage."

And in paragraph 304.d. that:

"The FAA design standard for pavements is based on a 20-year design life. The computer program [FAARFIELD] is capable of considering other design life time frames, but the use of a design life other than 20 years constitutes a deviation from FAA standards."

Under the current law covering construction funded by grants from the Airport Improvement Program (AIP) the statements above are referenced in paragraph 500 of the AIP Handbook (reference 2) by placing "funding limits on construction of pavement longer, wider, or stronger than specified in advisory circulars." The handbook goes on to say in paragraph 501 that: "The reconstruction, rehabilitation, pavement overlays, or major repairs of facilities and equipment are defined as eligible capital costs generally considered permanent with a 20-year life expectancy."

Materials and construction standards are contained in advisory circular AC 150/5370-10E "Standards for Specifying Construction of Airports." Therefore, AC 150/5320-6E and AC 150/5310-10E, combined with the AIP Handbook, require that the thickness, materials, and construction quality of a pavement be such that the pavement will last for a minimum of 20 years. It is required that this life be achieved without major rehabilitation activities and with the cost of planned maintenance activities borne by the airport owner. The stated assumption in AC 150/5320-6E is that the 20-year life applies to the structural capacity of the pavement and that easily repaired distresses, such as joint sealing in concrete pavements and minor cracking of the surface course in asphalt pavements, will be corrected by planned maintenance during the 20-year life period. More serious non-structural distresses, such as isolated slab failure in concrete pavements and severe rutting of the surface course in asphalt pavements, are not usually manifested within the 20-year life period. In fact, there is no time period specified in either of the two advisory circulars AC's) for anything other than structural life. The requirement in the AIP manual that the pavement last for 20 years implies, therefore, and considering that the ACs give the specifications under which AIP grants are administered, that the pavement's structural life must be 20 years, and that all other distresses that cause the pavement to fail to fulfill its functional requirements, and require more than planned maintenance to restore functionality, must not be manifested for at least 20 years. Even though there are no procedures specified in the ACs to explicitly design for a given time period based on anything other than structural life.

Since the plan described in this report is concerned, at least in part, with evaluating the feasibility of life extension and life cycle cost analysis, a distinction must be made between structural life and functional life. The pavement condition index (PCI) visual inspection methodology lumps all pavement distresses into a single number and is used primarily for planning maintenance and repair activities. This makes sense because structural failure is a form of functional failure, even though the specifications in the ACs make a clear distinction between the two and some degree of structural failure can be tolerated in a functional sense if properly managed, particularly for concrete pavements. Although not normally used in maintenance planning, the PCI number for a given pavement includes a subset of structural distresses called the structural condition index (SCI). SCI is used explicitly for the structural design of concrete pavements in the new FAA design procedure, but not for asphalt (flexible) pavements. In fact, the criteria for structural failure in asphalt pavements result in extremely conservative designs and full structural failure in the field, as defined by the criteria, is very rare.

> *...in evaluating the feasibility of life extension and life cycle cost analysis, a distinction must be made between structural life and functional life.*

In 2004, the FAA published a report, DOT/FAA/AR-04/46, *Operational Life of Airport Pavements*, (reference 3) that evaluated whether the application of FAA design and construction standards had resulted in the 20-year pavement life required by FAA technical and administrative standards. Existing condition survey data from FAA funded projects were acquired and the condition of both flexible (asphalt) and rigid (concrete) pavements were evaluated over time. Pavement condition was characterized with respect to measures of structural and overall performance, SCI and PCI, respectively, as previously discussed. Since, unlike rigid pavements, a recognized procedure for computing the SCI of flexible pavements did not, and still does not, exist, for comparison to rigid pavements, the SCI of flexible pavements was assumed to be comprised of structural distresses related to flexible pavements, e.g., rutting and fatigue cracking.

As shown in Figures 3 and 4, the report concluded that the structural and overall performance of both rigid and flexible pavements designed and constructed in accordance with FAA standards are comparable, with slightly better performance attributable to rigid pavements. The report also indicated that application of FAA standards, presumably combined with timely maintenance activities, will result in acceptable levels of functional (PCI) and structural (SCI) performance after 20 years. In this case, structural performance is represented by SCI and functional performance is represented by PCI minus SCI (the difference between the vertical bars in the figures). However, the existing data were insufficient to allow reliable extrapolations much beyond 20 years.

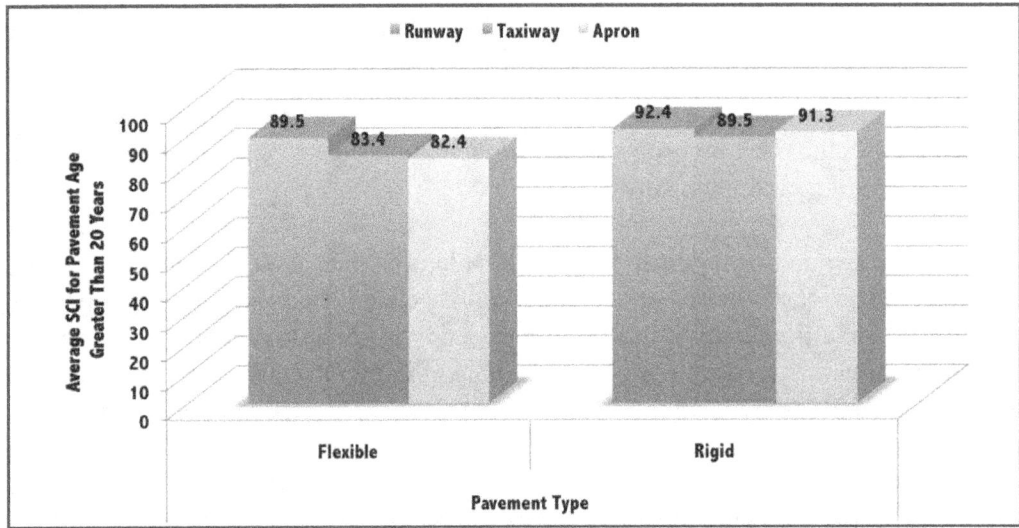

Figure 3: Average SCI for Pavements Older than 20 Years

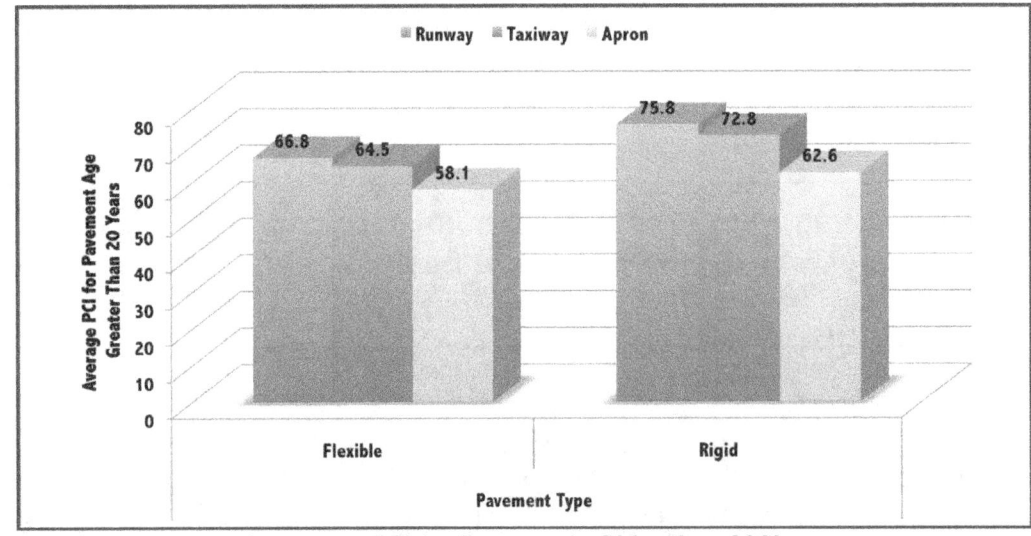

Figure 4: Average PCI for Pavements Older than 20 Years

> *...application of FAA standards, presumably combined with timely maintenance activities, will result in acceptable levels of functional (PCI) and structural (SCI) performance after 20 years. However, the existing data were insufficient to allow reliable extrapolations much beyond 20 years.*

The conclusions (reached in reference 3) were based on the structural condition of the pavements evaluated because the objective of the work was to determine the ability of the structural design procedures to result in a pavement structural life satisfying the 20-year requirement of AIP sponsored projects. For rigid pavements, the field and design procedure definitions of failure are for SCI be equal to 80. This corresponds to fifty percent of the slabs having one or more structural cracks, which is the same as the definition of rigid pavement failure in use in FAA design procedures before the SCI system was introduced. Nonetheless, industry criticism of the SCI 80 definition was expressed in Chapter 9 of reference 3, where it is pointed out that FAA concrete overlay design requires input of the SCI of the existing pavement and that the input SCI value for overlay design can be considerably lower than 80. So, the allowable condition of the pavement before overlay is inconsistent with the definition of failure. This criticism can be countered to some extent by saying that it is possible, and frequently done, to let the pavement deteriorate below an SCI of 80 but with increased frequency of condition surveys and with maintenance performed as required to maintain the functional performance of the pavement at an acceptable level. Industry criticism was also leveled at the definition of SCI for structural failure-in-the-field of flexible pavements adopted in reference 3 because it is not a traditional numerical measure of structural distress in flexible airport pavements. The flexible pavement SCI is composed of 100 minus the sum of the pavement condition deducts for rutting and alligator cracking. Failure was set at an SCI of 80. This is clearly not compatible with the full-scale test failure criteria, although it is difficult to see how the field and full-scale test criteria could be made compatible. Setting the full-scale test rutting criterion at, say, 1 inch would be extremely conservative and would not be an indication of structural failure. But allowing a busy runway to accumulate ruts deeper than 1 inch would be unsafe. Perhaps to put this discussion in a clearer perspective, the following are comments by Dick Ahlvin quoted in reference 4:

> *"Pavement failure is a conceived condition, which rarely, if ever, can be considered as a sudden occurrence or even a condition attained at a particular discernible time. Deterioration and development of distress occur over a period of time with continuing use. We speak of failure as a particular condition and we attempt to quantify some combination of attributes, which have themselves had to be quantified in some fashion, as a particular measure of failure. We need to do this and to continue to perfect the process, but we must not let this lead us to believe in a unique failure point or condition."*

In the same spirit, reference 3 states the following on page 6-7:

> *"However, in practice, the end of pavement life is determined by the airport operator, or by other external factors, and may occur when SCI is significantly lower than 80, or when PCI has reached a threshold value requiring remediation even though the SCI is still significantly above 80. In addition, threshold values are different for pavements with different functional purposes (runway, taxiway, or apron) and when the pavements are located in different environmental or geographical regions. The threshold value also depends on the airport size: large, medium, or small hub, or general aviation airport."*

…the end of pavement life is determined by the airport operator, or by other external factors, and may occur when SCI is significantly lower than 80, or when PCI has reached a threshold value requiring remediation even though the SCI is still significantly above 80.

Despite these obvious difficulties, structural failure criteria must still be defined for both full-scale test data, to calibrate the design procedures, and for failure-in-the-field, to identify incipient failure. It is possible that, in some cases, life-cycle cost analysis could provide a way to allow a range of the structural condition index to be incorporated in the design procedure, as discussed later.

When designing a pavement for structural failure the first step is to estimate the traffic the pavement is expected to experience over the design life and then to calculate the total number of departures for the complete traffic mix. Pavement thickness is therefore calculated as a function of traffic loading rather than time. Changing the design life then becomes a matter of extrapolating further into the future with respect to predicted traffic (wheel loads, gear configurations, frequency and changes in capacity) and, possibly, incorporating the effects of changes in the structural properties of the pavement. But, in principle, and assuming good quality construction with strict adherence to material specifications, the calculation of pavement thickness for an extended period of time is straightforward.

On the other hand, when designing, or specifying, for functional requirements, quantitative procedures are not available. Take, for example, the prediction of the life of a surface asphalt layer under the effects of two of the most common functional failures: asphalt rutting and groove deterioration. The procedure for asphalt surface mix design (P-401) currently does not include tests for predicting performance so, assuming that the current mix design is suitable for 20 years without exceeding the specification for functional failure due to rutting, it is impossible to predict how much longer than 20 years the pavement may last without excessive rutting, considering the changes in asphalt properties with time. The only guide is experience and engineering judgment. Future FAA specifications for asphalt mix design incorporating the Superpave methodology will almost certainly include laboratory tests for predicting rutting performance. But even then, correlation of the laboratory test results with field experience will be required before extrapolation over extended periods of time is possible. Groove deterioration is presumably related to the rutting performance of an asphalt mix, but quantification of the relationship is only possible through field experience. At present, anecdotal evidence suggests that a significant proportion of maintenance work on heavily trafficked asphalt pavements is remove and replace the top lift of the pavement after only about 10 years to restore the grooving to its specified geometry. If true, does this mean that three complete remove, replace, and re-groove operations would be needed on a pavement designed to have a structural life of 40 years? Apart from obscuring the rutting performance of the pavement, to rationally analyze and cost such operations and to perform a meaningful life cycle cost analysis would require verifiable information on the following:

1. Initial pavement design and construction methods.
2. Expected pavement performance over time.
3. Maintenance activities.
4. Itemized costs for the construction and maintenance work expected to be performed, including unit costs for materials procurement and placement in a standardized format.

But even then, if the groove rehabilitation turned out to become a significant proportion of the life-cycle cost over 40 years, improvements in mix design and grooving practices would be expected to drive innovation over such a long period of time and act to reduce the life-cycle costs compared to the initial estimates.

Similar arguments can be given for considering the effects of other forms of asphalt pavement deterioration, such as alligator cracking, weathering, and raveling, on the life-cycle cost of operating an asphalt pavement over an extended period of time. Remembering that, for life-cycle cost analyses to be useful in project design and selection, they need to be done as soon as the operational specifications of a project are available and before detail designs are prepared. This requires ready access to appropriate information gathered from previously completed projects. The central element in the plan described in this report is therefore to continue development of the pavement management database software application named FAA PAVEAIR and to populate at least one implementation of the software with information as itemized above. Other computer programs developed over the last 18 years for pavement evaluation will be integrated into the FAAPAVEAIR system and used for generating information on pavement performance in addition to that already contained in PCI.

> *... the central element in the plan is therefore to continue development of the pavement management database software application named FAA PAVEAIR and to populate at least one implementation of the software with information as itemized above.*

3.0 FAA PAVEAIR and Other Support Software:

FAA PAVEAIR is a web-based airport pavement management system that provides users with historic information about the construction, maintenance, and management of airport pavements. Also, the Maintenance and Repair (M&R) function of the program offers users a planning tool capable of modeling airport pavement surface degradation due to external effects such as traffic and the environment. The program has been developed for installation and use on a stand-alone personal computer, a private network, an intranet or the internet. An implementation of the internet version of FAA PAVEAIR is installed and supported on a server at the William J. Hughes Technical Center, and is accessible from the FAA PAVEAIR website, url: http://faapaveair.faa.gov/. It is expected that the database will be loaded with information from eligible FAA sponsored projects as the data becomes available.

Except for being a web-based application, the initial release of FAA PAVEAIR is functionally compatible with Micro Paver 5.3. Over time the program will be extended to satisfy the FAA's requirements with respect to classifying and quantifying the performance of civilian airports and with respect to extending pavement life for large hub airports. Among the functions expected to be added to the program are:

1. The ability to track the performance of individual distresses over time.
2. Additional distresses or independent measures of performance such as roughness.
3. Improved cost information entry and retrieval.
4. Improved prediction of remaining pavement life.
5. Provision for automatic calculation of life-cycle costs.
6. Integration, or interoperability, of other programs for independent design and evaluation of pavement items or measures of performance.
7. Network-wide analysis procedures.
8. Integration with geographic information systems (GIS).
9. Traffic data.
10. Environmental (climate) data.

Taken together, implementing all of these functions, and maybe more, will require a considerable amount of work. One of the first tasks to be undertaken under the plan will be to finalize the list, prioritize the order of implementation, and prepare a tentative schedule.

With regard to the integration of FAA PAVEAIR with other programs, figure 5 shows the currently available programs clustered with FAA PAVEAIR but indicating their generally subordinate role to FAA PAVEAIR and showing their primary function in pavement design and evaluation.

Each of the satellite programs is tied to a specific advisory circular as listed in Table 1. The current status of each of the programs and the work expected to be done under the plan is discussed in the following sections.

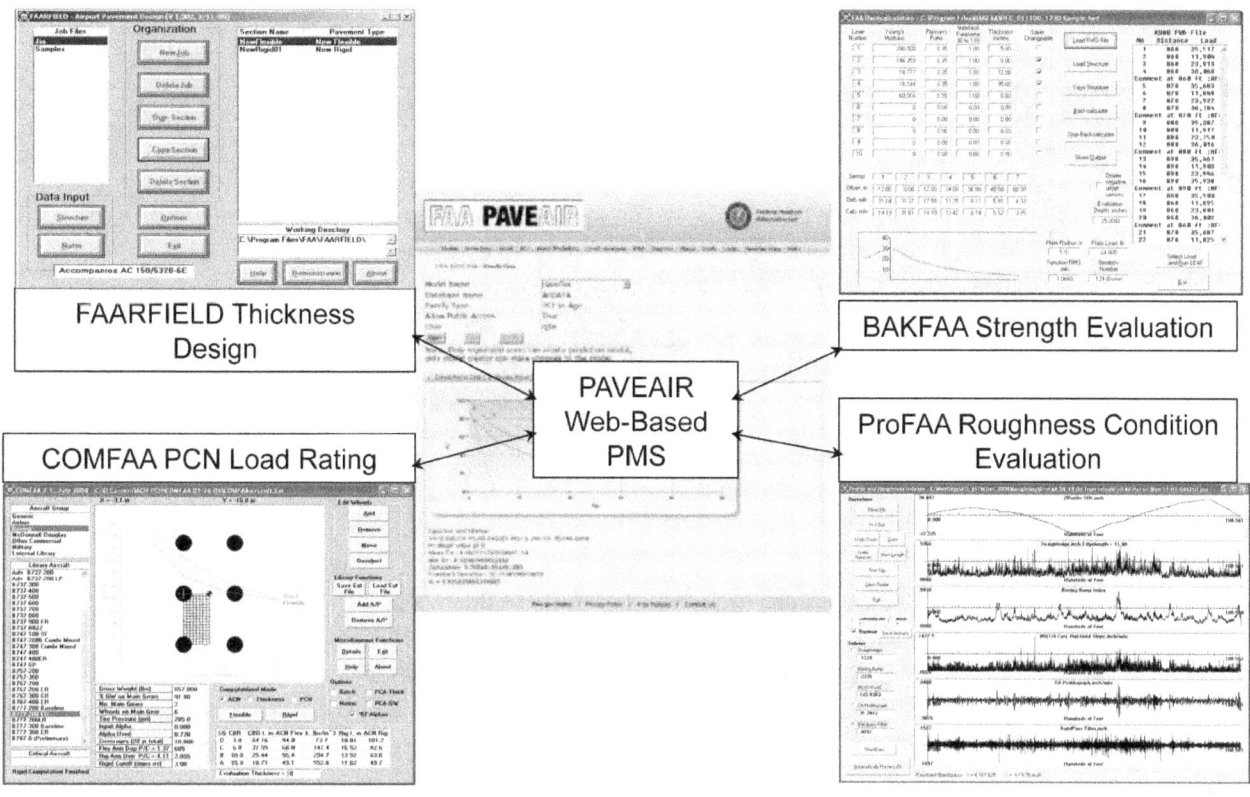

Figure 5: FAA PAVEAIR Integration: subordinate role of other programs and teir primary function.

Table 1. FAA Pavement design and evaluation programs developed under the FAA airport pavement R&D program.

Name	Date of Adoption	Advisory Circular	Description
PAVEAIR	2011	NA	Web-based application for airport pavement management, including PCI evaluations.
COMFAA 3.0	2011	150/5335-5B	Automatic PCN computation
FAARFIELD 1.3	2009	150/5320-6E	FAA Rigid and Flexible Interactive Layer Design. Fully tested thickness design. Uses NIKE3D for rigid and LEAF for flexible.
ProFAA	2099	150/5380-9	Longitudinal roughness profile analysis, roughness index computation, and aircraft ride simulation.
LEDFAA 1.3	2003	150/5320-6D Change 3	Rewrite of LEDFAA 1.2 as a 32-bit program. Uses LEAF instead of JULEA. Updated flexible failure model. Updated aircraft library, includes A380.
FEAFAA		NA	3D FEM program for rigid pavement response computation. Up to 9 slabs. Used to improve and extend FAA-NIKE3D.
FAA-NIKE3D		NA	3D finite element modeling system. Custom modification of the NIKE 3D FEM system developed by Lawrence Livermore Labs.

Name	Date of Adoption	Advisory Circular	Description
COMFAA 2.0	2006	150/5335-5A	ACN computation and thickness design by the FAA CBR and Westergaard methods.
BAKFAA	2003	Recommended in 150/5370-11A	FAA Backcalculation of elastic layer properties using LEAF. Also computation of elastic layered system responses and used for LEAF development.
LEAF	2003	NA	Layered Elastic Analysis FAA. Windows DLL layered elastic computational engine written in Visual Basic.
LEDFAA 1.2	1995	150/5320-16	FAA Layered Elastic Design. Windows-based 16-bit thickness design for flexible and rigid pavements. Used JULEA layered elastic computational engine.

3.1 FAARFIELD

FAARFIELD is the current FAA standard method for thickness design of airport pavements. It has been the primary focus of our R&D work to date, with the failure models of both the flexible and rigid design procedures being thoroughly calibrated with the addition of full-scale test data obtained from operation of the NAPTF. The rigid design procedure has additionally been considerably improved over both the previous Westergaard- and layered elastic-based procedures. Nevertheless, the program still requires further refinement and explicit consideration of additional failure modes. These requirements are discussed below, but the amount of work to be put into them will be much reduced, as a percentage of resource allocation, compared to previously.

The program is written in Microsoft Visual Basic and Intel Visual Fortran, and in most respects is considered to be a mature application. From a systems programming aspect, the program will receive normal maintenance, bug fixes, and adjustments to comply with changes in operating systems and computer systems' standards and protocols, except under the integration of the programs. From a technical programming aspect, refinements and improvements to the program are planned to include the following:

1. Add automatic computation of subgrade compaction requirements for output each time a thickness design is run. Analysis of candidate response criteria is underway. When the analysis is complete the criteria for implementation will be selected and incorporated in the program. Full-scale test and field data will be collected where possible and added to the current CBR-based

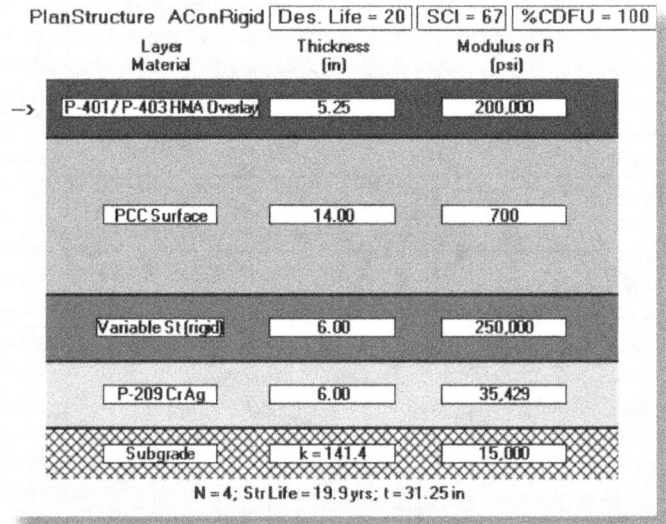

Figure 6: FAARFIELD structure for design of an asphalt overlay on rigid pavement.

verification data. The automatically computed compaction requirements will replace the tabulated requirements currently incorporated in AC 150/5320-6E.

2. Recalibrate the failure model in the flexible pavement design procedure to satisfy the following conditions.

    a. Match, to the extent possible, the effective layer equivalency factors in FAARFIELD with the accepted equivalency factors used in the previous CBR design procedure.

    b. Based on the results of the full-scale traffic tests from construction cycle 5 (CC5), exclude from the calibration the Boeing 747 thickness designs computed with 16 wheels by the CBR method.

    c. Match, to the extent possible, results from FAARFIELD with results from the CBR method obtained using the new ICAO method alpha factor curves.

3. Analyze the results of the construction cycle 6 (CC6) NAPTF traffic testing to determine the effect of concrete strength and stiffness on rigid pavement structural performance (SCI). Modify the rigid pavement design procedure in FAARFIELD if necessary.

4. Complete implementation of the energy-based model for hot mix asphalt cracking in the flexible design procedure and study the implications with respect to stabilized base layer cracking and the performance of perpetual pavement structures.

5. Improve the asphalt-on-rigid overlay design procedure based on results from the full-scale reflection cracking tests to be run at the NAPTF. A full-scale test rig is in process of construction. The rig consists of two 15 ft by 15 ft steel baths filled with refrigerated concrete and free floating on a sheet of Polytetrafluoroethylene (PTFE) in a specially constructed concrete basin. The steel baths are connected by hydraulic actuators which move the baths, and the contained concrete slabs, horizontally relative to each other. Asphalt overlays will be constructed on the top of the concrete and subjected to simulated environmental temperature expansion and contraction loading. The slabs will also be able to be loaded by the NAPTF test vehicle. The first tests are planned to be run in late 2011.

6. Refine the rigid pavement unbonded rigid-on-rigid overlay design procedure based on the results of IPRF Projects 04-2 and 06-3. The full-scale traffic testing for these projects was performed at the NAPTF under Construction Cycle 4 (CC4).

7. Develop a conceptual procedure for incorporating slab curling and top-down cracking in the rigid pavement design procedure. This may require a multiple-slab FEM model and multiple load cases to calculate critical stresses. Determine the expected computing requirements and, if feasible, begin program implementation.

3.2 COMFAA 2.0 and 3.0

COMFAA 2.0 is a computer program written in Microsoft Visual Basic 6.0 (VB6) to compute Aircraft Classification Numbers (ACN) by the standard ICAO method. COMFAA 3.0 is a VB6 computer program written to compute Pavement Classification Number (PCN) by the procedure recommended in AC 150/5335-6B. COMFAA 3.0 is considered to be a mature product, except for the integration of the programs. A decision will be made early in the plan as to whether to continue support of COMFAA 2.0.

3.3 BAKFAA

BAKFAA is a VB6 computer program written to backcalculate layer properties of a layered elastic pavement structure. The program is currently in the process of being converted to VB.NET and minor enhancements are being made to the user interface. The program is considered to be mature and is unlikely to see significant changes over the next 10 years.

3.4 ProFAA

ProFAA is a VB6 program written to analyze pavement longitudinal elevation profiles. Although mainly aimed at providing information related to ride quality and potential aircraft structural damage, two options related to new pavement acceptance and quality control are included. The first is a physical straightedge simulation and the second is a profilograph simulation with the capability of doing an automated bump analysis for identifying "must grind" areas. The pitch and bounce response of four representative aircraft can be simulated and peak acceleration and root mean square (rms) acceleration output for ride quality evaluation. A Boeing Bump analysis can be performed, with the output normalized to the Boeing threshold for excessive potential of structural damage being caused to an aircraft traveling on a runway pavement (reported as a "Boeing Bump Index" (BBI)). For reference, the highway standard International Roughness Index (IRI) is computed as well.

At present, the BBI is the only established standard for quantitatively identifying excessively rough airport pavements, and AC 150/5380-9, "Guidelines and Procedures for Measuring Airfield Pavement Roughness," provides guidelines for computing and applying the BBI. However, the BBI is strictly applicable only for aircraft traveling at close to the takeoff and landing speeds. A more general method of quantifying the ride quality of airport pavements is required that is applicable to both taxiway and runway surfaces and that can be used to track the roughness of a pavement over time in a pavement management database. Work is underway to develop such a roughness index using the FAA

Figure 7: The FAA 737 simulator operated by AFS-450 at The Mike Monroney Aeronautical Center, Oklahoma City. *(An interior photo appears on page 12.)*

full-motion B-737 flight simulator located at the Mike Monroney Aeronautical Center. It is expected that approximately four more years will be required to complete the development of procedures for computing the index using equivalent aircraft simulations as in ProFAA, at which time computation and output of the index will be implemented in ProFAA.

3.5 Software Integration

At present, the pavement design and evaluation computer programs described above are not compatible with each other to the extent that results produced by one program can be automatically transferred to another program. Copying, say, the PCN value produced by COMFAA into a field contained in the FAA PAVEAIR database can only be done manually. Nor can the aircraft entered for a given job in FAARFIELD be automatically transferred to COMFAA for computation of PCN values because the external file formats are different. As shown in figure 5 above, FAA PAVEAIR is considered to be the central, or unifying, element among the different pavement design and evaluation applications, and a case can be made for embedding the satellite applications directly into PAVEAIR. But this would make further development and future maintenance of the individual applications much more difficult than if they are kept as stand-alone applications. And, users would have to manage a single, very large application, without the ability to carry out the individual functions of, for example, thickness design and roughness profile analysis in stand-alone mode when desired. A gradual process of integration will therefore be followed, with the ultimate goal being to have FAA PAVEAIR capable of sending and receiving data from the FAA, and triggering operation of the functions in the other applications where the functions can be fully automated. The integration process is planned to be implemented with the following steps.

> *...integration of FAA PAVEAIR with FAARFIELD, COMFAA, BAKFAA, and ProFAA programs is essential. The ultimate goal is to have FAA PAVEAIR capable of sending data to other applications where the functions can be fully automated.*

1. Convert all of the programs to a common programming language and develop and maintain them under a common development environment. Specifically, Microsoft Visual Basic.NET and Microsoft Visual Studio.NET (currently at version 2010) or their replacements in the future Microsoft development cycles. The only exception will be the finite element subroutines in FAARFIELD, which will be continued in Fortran for as long as they can be maintained to be compatible with Visual Basic.
2. Convert all external data files to XML format with compatibility in the data structures between the individual applications.
3. Include a function in FAA PAVEAIR to enter and maintain an aircraft list associated with each pavement section or feature in the database. The list shall be capable of being passed to the other applications.

4. Provide functionality in FAA PAVEAIR to control the other programs without user input to the other programs. Unless future software technology is developed in the future, this may only be possible under restricted conditions if the independent operation of the other programs is to be maintained without developing and maintaining two versions of the same application, for example, one Windows-based and the other Web-based.

The first three steps are conceptually straightforward and are expected to be completed in approximately three years.

> *...the use of life cycle cost analysis (LCCA) for selection of airport pavement construction type and bidder selection appears to have become more popular in recent years, perhaps because of the increasing cost of all commodities, social pressure to maintain environmental standards, and concern about climate change.*

4.0 Life Cycle Cost Analysis (LCCA):

The use of life cycle cost analysis (LCCA) for selection of airport pavement construction type and bidder selection appears to have become more popular in recent years, perhaps because of the increasing cost of all commodities, social pressure to maintain environmental standards, and concern about climate change. The more traditional approach is to select the pavement type, perform a thickness design, and select the construction contractor based on the prices bid to construct the designed structure. Various combinations of how the steps are carried out are possible, but the central element in the costing and selection process is the thickness design. An important aspect of the FAA thickness design procedures is that the materials are carefully specified and the inputs for the thickness design are restricted to tightly controlled ranges or precise values. For example, in a standard FAARFIELD flexible design, the stiffnesses of the individual layers cannot be changed except for the subgrade. Similarly, for a rigid design, the concrete stiffness is fixed and the input for concrete strength is determined by well defined requirements in the advisory circulars. It is recognized that these constraints on input values can be somewhat unrealistic under some circumstances, but their use has become established over many years of experience and design procedure development. One reason for constraining the inputs is to prevent building overly weak pavements by incorrect selection of design inputs through inexperience or by mistake. Another reason is to prevent building overly strong pavements by inappropriate selection of design inputs. The first case is clearly undesirable because the pavement is likely to fail prematurely and require additional expenditure of funds for rehabilitation. But the second case is also undesirable because, without a special waiver, it does not comply with the AIP capital expenditure requirements and will lead to higher initial costs than necessary.

Now, with LCCA added to the design and selection process, significant deviations from the desired cost and expenditure structure over time are more likely than with the traditional approach. Current FAA guidelines for the use of LCCA in pavement design and construction consist of four pages in Appendix 1 plus one paragraph in the body of AC 150/5320-6E and one page in the AIP Handbook. Also, a report has been published recently by the Airfield Asphalt Pavement Technology Program (AAPTP) (reference 5) which discusses the application of LCCA to airport pavements together with a software application for running analyses automatically. But this is a research report and does not constitute a standard or regulation. The report, through its literature search and case studies shows that the amount of publically available information on this topic is small. These factors argue strongly for a comprehensive study of LCCA procedures in airport pavement design and construction and an examination of the need for more comprehensive and stronger standards controlling the application of LCCA procedures, perhaps with tighter coupling to the materials and thickness design standards.

> *... one reason for constraining the inputs is to prevent building overly weak pavements, another reason is to prevent building overly strong pavements. The first case is clearly undesirable and the second case is also undesirable because, without a special waiver, it does not comply with the AIP capital expenditure requirements and will lead to higher initial costs than necessary.*

4.1 Materials and Construction Costs

As the FAA develops design software for runways and taxiways, the inclusion of life cycle cost analysis (LCCA) software for pavement design and management is a logical step. Standardization of the LCCA procedure will help the FAA and airport owners better evaluate construction and maintenance alternatives. However, having accurate cost estimates is critical to the outputs of the LCCA.

> *... an important aspect of the FAA thickness design procedures is that the materials are carefully specified and the inputs for the thickness design are restricted to tightly controlled ranges or precise values.*

Under the Airport Cooperative Research Program (ACRP), a project has been initiated to develop a user friendly, interactive, construction cost estimating model and associated database. (reference 7) There are a number of commercial construction cost databases that can be subscribed to by contractors and designers for estimating construction materials and labor. The ACRP project will most likely synthesize cost information from various commercially available databases to produce improved airport construction specific cost models. The resulting model and database must have the ability to be adjusted by the user to reflect local contractor and material supplier markets as well as localized adjustments for labor costs, especially as the rise in construction cost has outpaced inflation for a number of years. The high volatility of commodities such as asphalt binder and cement are very region specific as there are large transportation costs associated with these materials.

> *...since the publicly available information on LCAA is scarce, a comprehensive study of LCCA procedures in airport pavement design and construction and an examination of the need for more comprehensive and stronger standards controlling the application of LCCA procedures, perhaps with tighter coupling to the materials and thickness design standards is needed.*

The combination of the AAPTP LCCA software and the ACRP Cost estimating model and database can be a valuable tool for the FAA and airport owners when evaluating project alternatives. If the ACRP cost estimating model and database are successfully produced, the database must be maintained for it to be continually accurate and useful. Right now, databases are maintained by private industry via fees collected from users of the database. If the FAA is to include a cost tracking database there must be a mechanism in place to maintain and populate the database at no cost to the user. A standardization of the LCCA procedures and a common database will allow better comparisons of project alternatives when determining the best use of AIP funds. If the ACRP database is successful, the FAA should include a cost estimating model in conjunction with improved LCCA procedures in future pavement construction and maintenance software products. The FAA could use the ACRP database or further refine it for pavement specific projects.

4.2 Sustainability

There are many ways to define sustainability. Listed below are several examples:

"Development that meets the needs of the present without compromising the ability of future generations to meet their own needs" - Brundtland Commission 1983

"One that meets transportation and other needs of the present without Compromising the ability of future generations to meet their needs" - TRB 2005

" A holistic approach to managing an airport so as to ensure the integrity of the Economic viability, Operational efficiency, Natural resource conservation and Social responsibility (EONS) of the airport." - ACI-NA

No matter how sustainability is defined it still embraces what is called the triple bottom line, as shown in Figure 8. As airports embark on a sustainability programs, it is vital for each airport to establish its specific definition of sustainability to initiate the basis for future planning and implementation. An airport's definition of sustainability should relate to its unique circumstances and role within its community and environment.

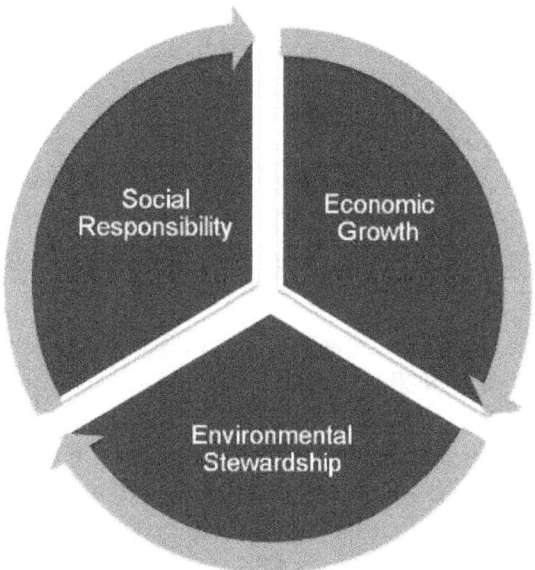

Figure 8: Triple-Bottom Line

> *... as airports embark on a sustainability programs, it is vital for each airport to establish its specific definition of sustainability to initiate the basis for future planning and implementation.*

Airports across the country and around the world have embraced sustainability and are implementing programs or initiatives at their facilities. Along with benefiting their communities and the environment, airports are finding that sustainability makes good business sense. A sustainability program often results in numerous benefits to the airport.

These include:
- Reduced Operating Costs
- Greater Utilization of Assets
- Reduced Environmental Footprint
- Operational Flexibility
- Enhanced Customer Service
- Optimization of New and Better Technologies
- Lowering Costs of Asset Development
- Integrated Design as a Way of Doing Business
- Improved Bond Rating
- Improved Benefits to the Community
- Enhanced "Green Design" criteria that integrate comprehensive and long-term environmental and economic benefits (reference 8)

Within the airport context, sustainability has broad implications throughout the entire system, including energy consumption; environmental impacts; and overall facility life-cycle costs.

> ... *sustainability has broad implications throughout the entire system, including energy consumption, environmental impacts, and overall facility life-cycle costs.*

Currently there are no nationally recognized sustainable rating systems for airfield pavements or other flat work in the aviation industry such as the Leadership in Energy and Environmental Design (LEED®). Many airports have adapted the LEED program and created airport-specific sustainability guidelines and metrics for their particular programs and have made their guidelines available publicly for all airports to utilize. Just recently, the Institute for Sustainable Infrastructure was founded in February 2011 through contributions by the American Council of Engineering Companies (ACEC), the American Public Works Association (APWA) and the American Society of Civil Engineers (ASCE). The Institute for Sustainable Infrastructure (ISI) is dedicated to working with public and private owners, designers, contractors, materials and equipment manufacturers, and suppliers to forge a new vision for sustainable infrastructure throughout North America. They are developing a new rating system called the envision™ sustainability rating system.

> ...*a new vision for sustainable infrastructure throughout North America, called the "invision" could be used to get "green" credits for the construction of airfield pavements.*

The construction of airfield pavements is a great opportunity for airports to be credited with the "green" credits often used in the above mentioned rating systems. As examples, these credits can be awarded for planning and constructing a balanced cut and fill pavement project to the use of recycled concrete aggregates as base material under airfield shoulders. To understand how pavements and the construction of them are currently being used, a review of current sustainability plans needs to be undertaken, as well as a review of the rating system being used at these airports and the system being proposed by the ISI. As part of the review and use of airport sustainability the issue of Life Cycle Assessment (LCA) of airfield pavements shall be investigated as to its feasibility and future effects, if any. These reviews will provide an outline on how airfield pavements are being rated and what practices are currently being used and possible future trends in pavement sustainability.

4.3 Integration with the Thickness Design Procedures

The rigid pavement thickness design procedure implemented in FAARFIELD is based on the SCI concept with, as previously noted, SCI being a subset of PCI and therefore readily available from FAA PAVEAIR for any pavement section actively maintained in a database. Although the current failure criterion is for an SCI of 80, rigid pavements are frequently operated beyond this level of structural deterioration, but with an expected increase in frequency of maintenance activities, and perhaps with changes in the type of maintenance or rehabilitation. Carrying this strategy one step further, the design procedure could be modified to allow different SCI values to be specified for failure and, when combined with different maintenance strategies, incorporated in the LCCA to find an optimum SCI for "failure" to give lowest life cycle cost."

> *...the rigid pavement design procedure could be modified to allow different SCI values to be specified for failure and, when combined with different maintenance strategies, incorporated in the LCCA to find an optimum SCI for "failure" to give lowest life cycle cost."*

The flexible pavement thickness design procedure in FAARFIELD is currently formulated for a structural life of 20 years. Pavements designed for heavy aircraft, as would be expected on a large hub runway or taxiway, have a standard structural requirement of five inches of surface course, a sufficient thickness of stabilized base course (asphalt or lean concrete), and a well-graded crushed aggregate subbase course sufficiently thick to protect the subgrade from rutting failure caused by shear flow. These structures are designed very conservatively, but the surface may fail functionally in less than 20 years, for reasons discussed above. Successive overlays add to the structural capacity of the pavement, or remove and replace operations maintain the structural capacity in its design condition. And it usually happens that the subgrade, base, and subbase layers never suffer structural failure as defined in the full-scale test methods used to calibrate the design procedure. A variant of the conventional flexible pavement structure, known as a perpetual asphalt pavement, aims to allow the structure to be designed for essentially unlimited surface remove and replace operations in a rational manner rather than as an ad hoc byproduct of the conventional design methodology. The general idea is to provide a rut resistant surface asphalt layer on top of a fatigue resistant asphalt base layer. For highway pavements a base layer can be full-depth to the top of the subgrade or the sub-surface layers can be some combination of different asphalt mixes and aggregate material (see reference 6). But for heavily loaded airport pavements it is likely that the presence of one or more aggregate layers would be needed for the most cost effective design. The thickness of the structural layers is generally determined so that the maximum horizontal strain in the fatigue resistant asphalt layers is held below the endurance limit for the particular mix design; hence the use of the term "perpetual." But this feature of the intended performance of the pavement makes it difficult to assign a value to the structural life for cost-benefit and life cycle cost analysis. Presumably, though, the true structural life of the pavement would be determined by deterioration of the asphalt over time, or by the inevitable, but difficult to predict, increase in traffic loading over time.

> *... a variant of the conventional flexible pavement structure, known as a perpetual asphalt pavement, aims to allow the structure to be designed for essentially unlimited surface remove and replace operations in a rational manner rather than as an ad hoc byproduct of the conventional design methodology.*

Implementing the so-far proposed and implemented perpetual pavement design procedures for highway pavements into FAARFIELD would be relatively straight forward except that suitable failure models for heavy airport pavements would have to be developed from full-scale test data. Selection of the analysis period will be the most difficult aspect of running LCCA on perpetual pavement designs together with selecting the frequency of overlay operations. Nevertheless, integrating perpetual pavement design methodologies and performance characteristics into LCCA, together with an expanded rigid design methodology, will be an important aspect of the plan.

> *... integrating perpetual pavement design methodologies and performance characteristics into LCCA, together with an expanded rigid design methodology, will be an important aspect of the plan.*

4.4 LCCA Software Integration with PAVEAIR

Reference 5 has a companion software package called AirCost. AirCost is a Microsoft Excel application with the LCCA functions written in Visual Basic for Applications. Deterministic and probabilistic analyses can be run independently. If a LCCA is to be run on a pavement for which data has been entered into PAVEAIR, it would be convenient if the data could be transferred automatically to AirCost or if the functions of AirCost were directly added to the PAVEAIR application. Adopting either of these two strategies would have the effect of forcing the extension of PAVEAIR to accommodate a realistic LCCA methodology in addition to the existing maintenance and repair functions. Then, when it is time to implement the findings and procedures of a broader-based LCCA project, PAVEAIR will have the basic functionality in-place. If extended pavement life strategies are eventually adopted, the LCCA and M&R functions could be merged and, as the pavement ages, the original LCCA maintenance projections could be updated periodically as pavement performance data is added to the database.

5.0 Non-Destructive Testing (NDT):

Due to the increasingly limited availability of access to airport pavement because of operational and maintenance constraints, a method for accurately assessing pavement condition in a minimum amount of time will continue to be a pavement management necessity. This pavement evaluation will be required to take place in off hours for airports such as over night. Therefore, this technology will require high-speed, accurate, and repeatable data acquisition without the benefit of sunlight.

> *... a method for accurately assessing pavement condition in a minimum amount of time will continue to be a pavement management necessity.*

One of the primary Non-Destructive Testing (NDT) initiatives in the future will be the acquisition of crack detection and pavement evaluation data from a moving vehicle. This will require the capability of crack identification, recording the crack dimensions, and the crack severity. This equipment will require Global Positioning to quickly locate areas of interest from the data acquisition and provide automatic notice to the reviewer where and when anomalies are detected. A Flight Line van has been purchased has had pavement imaging equipment for airport pavement evaluation istalled. The vehicle also includes Global Positioning and air-coupled Ground Penetrating Radar as part of the initial outfitting. The FAA envisions that the van will be used primarily to assess the condition of in-service airport pavement. Airport pavement data acquisition for new construction and in-service pavement is for research only.

Other pavement NDT initiatives include researching the viability of applying nanotechnology to pavement load response and evaluation, development of a procedure to use the Portable Seismic Pavement Analyzer (PSPA) as an evaluation tool for rapidly estimating the in situ strength of concrete pavements, evaluating the use of Magnetic Resonating Imaging (MRI) technology to verify dowel bar alignment and placement in rigid pavement, and assessment of groove performance with respect to trafficking in rigid and flexible pavement. The groove evaluation in flexible pavement will include the assessment of groove performance in heated pavement with respect to traffic and temperature.

5.1 Pavement Profiling and Grooving Measurement

In support of the ProFAA pavement roughness evaluation computer program, a portable inertial profiler was developed specifically for measuring longitudinal profiles of airport pavements. The profiler has been used over a period of more than 10 years to measure and compile a set of runway and taxiway profiles covering a wide range of pavement structures and levels of deterioration, from brand new to excessively rough. The profiles have been posted on the AJP-6310 website for download by interested parties and for use with ProFAA. A selection from the full set of profiles is also being used in the previously discussed B-737 simulator ride quality study.

The functionality of the profiler has recently been extended to allow the measurement of profiles with enough fidelity to measure the geometry of transverse grooves on runways. A separate computer program, ProGroove, has been developed to automatically detect individual grooves and to estimate the depth, width, and spacing of the grooves detected. The equipment and computer program will be further developed for use in new grooving acceptance tests and for periodic measurement of the grooving of in-use runway pavements for compliance with groove geometry standards.

6.0 New Materials and Processes

Each day new materials are being proposed for pavement construction from the use of nanotechnology in concrete mix design to the use of asphalt cement from shingles in hot mix asphalt. The difficulty in these new materials and processes are whether they are as durable under environmental and aircraft loading over a 20 or even 40 year design life then the current specified materials.

The first step is to determine which products and processes are currently being used in the industry that are not being used under the P-401 or P-501 specifications which show promise in their use in airfield pavements. Currently, some of the newer materials which are candidates for study and testing under this plan include:

- Polymer modified asphalt and binders (PMA and PMB).
- Warm mix asphalt (WMA).
- Geosynthetics, geogrids, and geotextiles.
- Stone matrix asphalts.
- Add-mixtures to concrete pavements
- Marginal quality aggregates for unbound base and subbase courses.
- Intelligent Compaction.

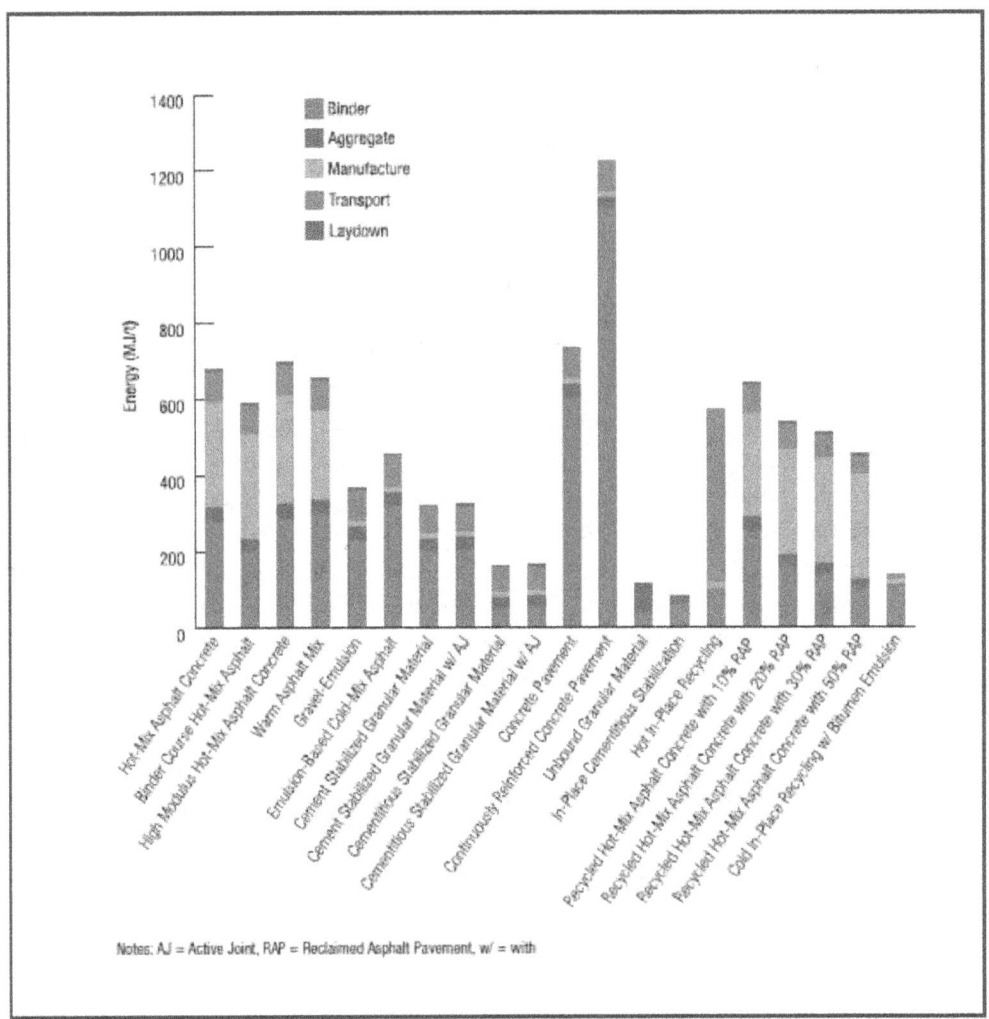

Figure 9: Construction Alternatives

Durability of pavements using recycled materials and alternate construction processes such as those shown in Figure 9 should be investigated and reviewed as to their feasibility on airfield pavements, whether it be at general aviation or large commercial hubs. Some alternatives may not be feasible for large commercial airports but work very well with GA airports. The assessments need to be determined through research and review of the current and future trends within the industry.

7.0 National Airport Pavement Test Facility (NAPTF)

Background on the NAPTF, and its role in past work, has been covered in the introduction has been reported in numerous publications, some of which are cited in the introduction. Therefore, this type of material will not be repeated here. Major projects which have been completed at the facility since its inception are listed below, where CC stands for "Construction Cycle."

1. CC1: Comparative study of the effect of four-wheel versus six-wheel loading on rigid and flexible pavement life under traffic loading on three subgrade strengths.
2. CC2 Test Strips: Study of the effects of construction techniques, materials, and slab geometry on the curling and traffic life performance of rigid pavements. The results were used to establish the testing methodology used in all subsequent rigid pavement testing.
3. CC2: Comparative study of the effect of four-wheel versus six-wheel loading on rigid pavement life under traffic loading on a medium-strength subgrade (about 8 CBR) on three different support conditions (on-grade, aggregate subbase, and econocrete subbase). The results were used to develop the rigid pavement design failure models in LEDFAA 1.3 and FAARFIELD.
4. CC3: Comparative study of the effect of four-wheel versus six-wheel loading on flexible pavement life under traffic loading on a low-strength subgrade (about 3 CBR) with four different pavement structure thicknesses ranging from very weak to very strong. The results were used to develop new Alpha Factors for use in ACN computations and to update the LEDFAA 1.3 and FAARFIELD flexible pavement failure models.
5. CC4: Study of four- and six-wheel loading on unbonded concrete overlays on concrete underlays on a single medium-strength subgrade with three different overlay thicknesses. Design, construction, test planning, and data analysis were performed by Quality Engineering Solutions (QES) under an Innovative Pavement Research Foundation project. Traffic testing and pavement response data collection were performed by FAA personnel at the NAPTF. Phases of the project included:
    a. Construct a pavement with new underlay and new overlay.
    b. Traffic the pavement to failure.
    c. Remove the overlay and replace with a new overlay on the deteriorated underlay.
    d. Traffic the pavement to failure
    e. Analyze the data and prepare a report, by QES.
    f. Further analysis of the data and incorporation of the results into FAARFIELD is expected in the future under FAA funding.
6. CC5: Study of the effects of gear interaction on flexible pavement life under traffic loading on a low-strength subgrade with three different pavement structures. Gear interaction was simulated by running a six-wheel gear configuration on one side of the pavement and adding four extra wheels to the six-wheel group on the other side of the pavement. Therefore comparing six-wheel

loading to 10-wheel loading. The results demonstrated that the layered elastic design model is more representative of full-scale test results than the CBR with Alpha Factors for wheel groups containing more than 6 wheels.

7. CC6: Study of the effects of concrete flexural strength on rigid pavement life under four-wheel traffic loading with two different subbase materials, econocrete and asphaltic concrete. Traffic testing on CC6 has just commenced and results from the tests are not yet available.

8. High tire pressure (HTP) testing: The performance of asphalt pavements under traffic loading is difficult at the NAPTF because the asphalt temperature cannot normally be controlled. However, a need arose to study the rutting performance of asphalt under high pressure aircraft tire loading and a 100-foot-long strip of pavement was constructed with an econocrete base layer having hydronic heating coils. Test conditions included: different tire pressures, different wheel loads, and different asphalt mix design specifications. Together with independent Airbus test results, the results of the HTP testing were used as supporting data in a successful request to ICAO to increase the maximum allowable tire pressures on asphalt surfaces at airports subject to ICAO standards. Future full-scale test work on asphalt surfaces is expected to be done on a specially constructed test machine described below.

Continued operation of the test facility is expected in the future in support of other projects and initiatives.

Current work includes:

1. Comprehensive posttraffic testing of the CC5 flexible pavement test items. Among other things, the results from this work will be used in the modeling of compaction requirements for flexible pavement construction.
2. Planning for CC7, which will include a study of the effects of overload on the life of flexible and rigid pavements.
3. Study of the effects of deicing chemicals on the cracking and deterioration of concrete slabs due to aggregate-silica reactivity (ASR) reactions.
4. Construction of a test rig for studying reflection cracking in asphalt overlays on concrete pavements. The rig is full-scale and incorporates two 15-foot square slabs so that a full-size asphalt paving machine can be used to place the asphalt. The representation of reflection cracking in the FAARFIELD design procedures is purely empirical and a better understanding of the underlying mechanisms is badly needed. Results from the reflection cracking rig will be used in conjunction with the theoretical modeling which has been underway at the University of Illinois for more than 10 years.
5. Study of the performance of standard and trapezoidal grooves in asphalt and concrete surfaces.

8.0 Field Testing

Full-scale testing at the NAPTF (indoor facility) has concentrated on load related effects on pavement failure. Environmental factors, coupled with traffic loads, play a significant role in pavement performance. Long-term field instrumentation studies were started by the FAA to collect pavement response and performance data under varied climatic conditions. The pavement response measurements will typically be used to investigate the behavior of a pavement with regard to a specific aspect of pavement performance (such as cracking of longitudinal construction joints in asphalt layers) or with regard

Figure 10: Site Installation at JFK

to a specific aspect of design modeling (such as concrete slab curling). Using results from these studies will improve the FAA's design procedures by including climatic effects on pavement behavior (such as slab curling, thermal gradients in Hot Mix Asphat (HMA) pavements, etc.). The FAA currently has instrumented pavements at Denver International Airport, John F. Kennedy International Airport, LaGuardia International Airport, and Atlanta Hartsfield-Jackson International Airport.

> ***...long-term field instrumented pavements at Denver International Airport, John F. Kennedy International Airport, LaGuardia International Airport, and Atlanta Hartsfield-Jackson International Airport will continue to provide information for pavement design.***

Laboratory testing results on pavement materials provides inputs for pavement design. However, the properties in the field might be different because of construction techniques and other in situ conditions. Therefore, determination of in situ material properties is very important to predict pavement life. These in situ tests must be simple, quick and yet measure index properties of soils and unbound materials. These properties shall include shear strength, resilient modulus, moisture content, density, etc. The application potential of in situ testing equipment such as LWD, vane shear, PSPA, DPSPA, BCD, etc., must be studied for different materials under different conditions. Correlations shall then be developed between different material properties.

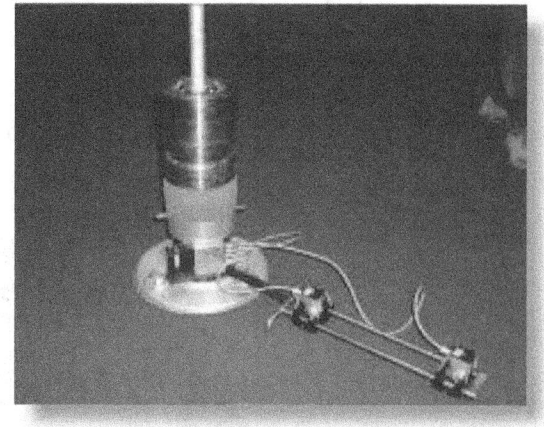

Figure 11: Lightweight Deflectometer (LWD)

9.0 High Temperature Pavement Test Facility (HTPTF)

The trend in the aircraft industry is to produce aircraft with extended range capability, which results in high gross weights and tire pressures. The effects of high tire pressure are localized and concentrated in the surface layers (like HMA). This makes it imperative to study the effects of high tire pressures on the HMA surface and also develop HMA mix design procedures to produce mixes that can withstand these anticipated high tire pressures. The work to-date for high tire pressures has primarily been laboratory tests and three cycles of full-scale testing on a limited area of heated test pavement at the NAPTF. Comprehensive full-scale tests at high HMA temperatures are needed so that the performance prediction models for HMA from laboratory tests can be validated and calibrated to the in situ pavement results. In the projects related to pavement material properties and surface layers, wheel load and tire pressures in combination with surface temperature are more critical than the gear load (due to minimum wheel load interaction effects). The Airport Pavement Test Vehicle (APTV) will be used to study the performance of different surface layer mixes. Pavement test lanes will be narrower for the APTV compared to the test lanes for the existing NAPTF test vehicle. It will be easier and more economical to insulate and heat the test pavement under the APTV.

> *... Comprehensive full-scale tests at high HMA temperatures are needed so that the performance prediction models for HMA from laboratory tests can be validated and calibrated to the in situ pavement results. A facility will be constructed to house the APTV to protect it from the weather elements and to provide a controlled environment for testing.*

A facility will be constructed to house the APTV to protect it from the weather elements and to provide a controlled environment for testing. The facility will be a steel frame tensioned fabric structure 300-feet long by 150-feet wide. Six test pavements 245-feet long by 20-feet wide will be constructed. Future research using the APTV will involve testing greener technologies such as Warm Mix Asphalt (WMA), Stone Matrix Asphalt (SMA), Asphalt overlays of PCC pavements, Recycled Asphalt Pavement (RAP),

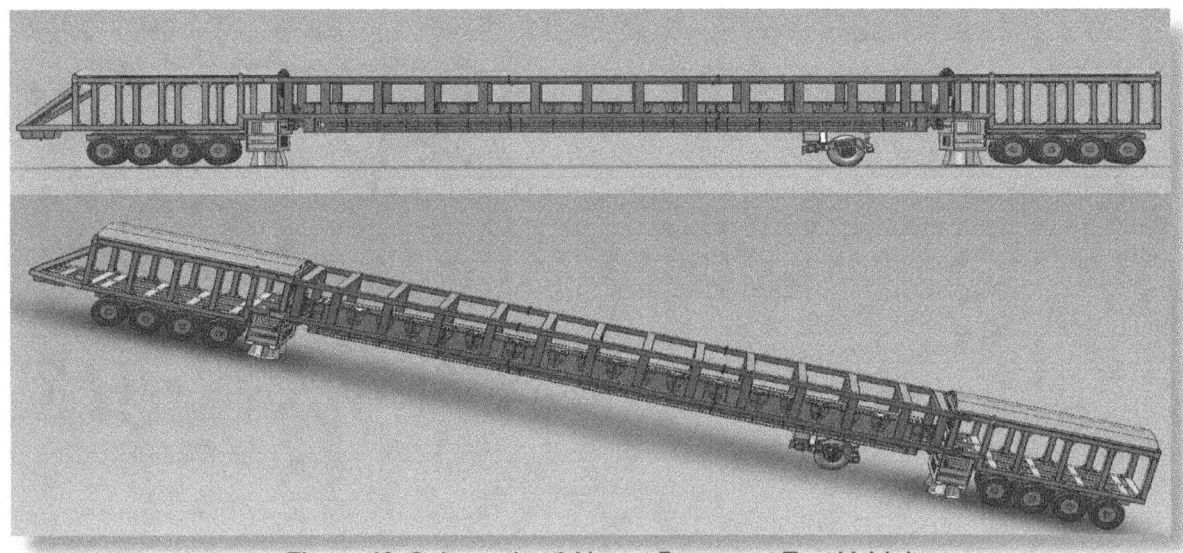

Figure 12: Schematic of Airport Pavement Test Vehicle

Polymer Modified Binders, and to study the shear failure of HMA. Multiple test lanes using these improved paving materials for surface layers and standard P401 mixes placed over a high-strength subgrade and with strong base and subbase will be constructed. The test lanes will be tested using APTV. HMA cores extracted from the test lanes will be tested in the laboratory using the FAA's modified Asphalt Pavement Analyzer (APA). Performance of these layers will be compared with the performance of standard P401. Test data will be used to develop rutting and fatigue cracking models for improved materials. These will ultimately lead to the development of material and construction standards and specifications for use on airport pavements.

Another aspect of APTV research will be to improve and develop new and innovative pavement instrumentation techniques. The application potential of microelectronic monitoring systems (MEMS), fiber-optic sensors, and nanotechnology based sensors will be studied.

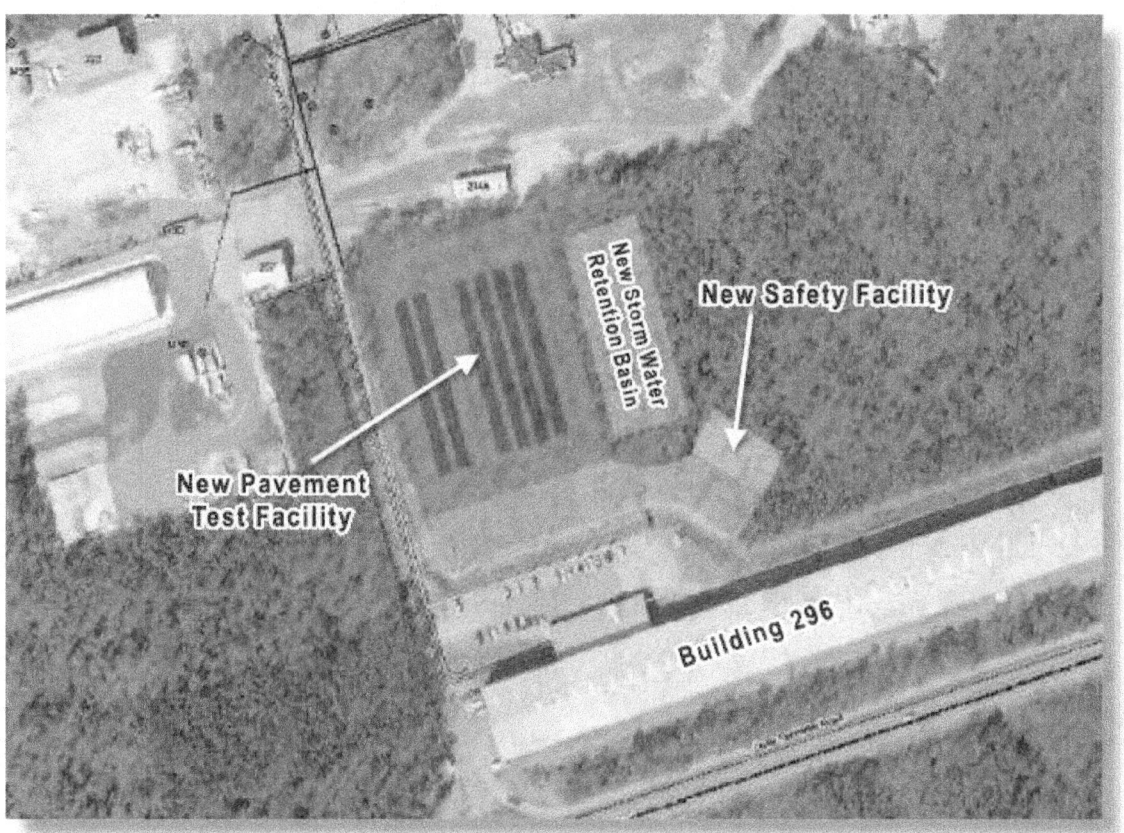

Figure 13: APTV Test Facility Layout

References:

1. Research Directorate for Airport Technology, Airport Technology Branch, "Airport Pavements – Solutions for Tomorrow's Aircraft," U.S. Department of Transportation, Federal Aviation Administration, FAA Technical Center, NJ (available for download at http://www.airporttech.tc.faa.gov/naptf/download/index1.asp).
2. U.S. Department of Transportation, Federal Aviation Administration, Order 5100.38C, SUBJ: Airport Improvement Program Handbook, June 28, 2005.
3. Garg, Navneet, Guo, Edward H., McQueen, Roy D., "Operational Life of Airport Pavements," Report DOT/FAA/AR-04/46, Federal Aviation Administration, December, 2004.
4. Chou, Yu T., "Investigation of the FAA Overlay Design Procedures for Rigid Pavements," Report DOT/FAA/PM-83/22, Federal Aviation Administration, August 1983.
5. Applied Research Associates, "Life Cycle Cost Analysis for Airport Pavements," Final Report on AAPTP Project 06-06, January, 2011.
6. Newcomb, David, E., Willis, Richard, Timm, David, M., "Perpetual Asphalt Pavements: A Synthesis," Asphalt Pavement Alliance, Publication IM-40, available on AsphaltRoads.org.
7. Airport Cooperative Research Program (ACRP) Project 01-19 – "Airport Capital Improvements: developing a Cost0Estimating Model and Database" from website http://www.trb.org/ACRP/Public/ACRP.aspx
8. "Sustainable Aviation Resource Guide", Sustainable Aviation Guidance Alliance (SAGA) Website: http://www.airportsustainability.org
9. Road rehabilitation Energy Reduction Guide for Canadian Road Builders, Natural Resources Canada, 2005

Airport Technology Research Plan
Milestones and Costs

Airport Pavement R&D Project Milestones

Fiscal Year	12	13	14	15	16	17	18	19	20	21
Advances in FAA PAVEAIR ($25M)										
Database of materials and construction costs				▒	▒	▒				
Life Cycle Cost Analysis: database, LCCA procedures, structure optimization	▒	▒	▒	▒						
Integrate design and evaluation software as web-based applications		▒	▒	▒	▒					
Advances in FAARFIELD Design Procedures ($20M)										
Subgrade characterization model and design-based compaction criteria	▒	▒								
Flexible pavement subgrade failure and asphalt fatigue models	▒	▒								
Rigid pavement top-down cracking failure model	▒	▒	▒	▒						
Reflection cracking model for asphalt overlays		▒	▒	▒	▒					
40 Year Design Life	▒	▒	▒	▒	▒	▒	▒			
Pavement Surface Geometry Studies ($15M)										
Ride quality criteria in simulator studies	▒	▒	▒							
Comparison of full-scale aircraft response with simulator response	▒	▒	▒	▒	▒	▒				
Runway intersection grading criteria			▒	▒	▒					
Automated groove inspection software			▒	▒	▒					
Non Destructive Testing (NDT) ($22M)										
Evaluation of NDT applications in the NDT Vehicle	▒	▒	▒	▒	▒	▒	▒	▒	▒	▒
Automated crack detection and condition index determination software	▒	▒	▒	▒	▒	▒	▒	▒	▒	▒
Concrete acceptance testing procedure with PSPA	▒	▒	▒							
Overload Criteria ($8M)										
Full-scale testing at the NAPTF				▒	▒	▒				
Overload criteria for flexible and rigid pavements			▒	▒	▒	▒	▒			
Alternative ACN-PCN procedures		▒	▒	▒	▒	▒	▒			
National Airport Pavement Test Facility ($15M)										
Support for theoretical studies and laboratory-based testing	▒	▒	▒	▒	▒	▒	▒	▒	▒	▒
High Temperature Pavement Test Facility ($12M)										
Rutting criteria for high pressure tires and alternative materials				▒	▒	▒	▒			
Base and subbase overload and rutting criteria				▒	▒	▒	▒			
Field Testing ($8M)										
In-use pavement load and environmental response characteristics	▒	▒	▒	▒	▒	▒	▒	▒	▒	▒

Airport Safety R&D Project Milestones

Fiscal Year	12	13	14	15	16	17	18	19	20	21

Winter Operations: ($23M)
Heated Pavements, Braking Performance

- Instrumented heated pavement evaluation at the NAPTF
- Engineering study of a hybrid heated pavement system
- Development of brake system control in ASBS Simulation Laboratory
- Integration of developed brake control system into FAA Boeing 727
- Development of mathematical models that represent ASBS

Next Generation Aircraft Rescue and Firefighting: ($22M)
Cargo Aircraft, Advanced Composite Materials, 2nd Level Access

- Fire testing of cargo aircraft interior lining material
- Effectiveness of aircraft skin penetrating nozzles on interior cargo fires
- Development of guidance for tactics and agent requirements for cargo fires
- Evaluation of tools, tactics, and techniques for composite fires
- Hazards associated with composite fires; Investigation of Biofuels
- Investigation of incidents involving interior fires
- Development of design concept for interior access vehicle

Visual Guidance: Safety, Capacity and Environment ($32M)

- Development of new heliport lighting
- Identification of LCGS framework specifics
- Advanced Taxiway Guidance System Improvements
- Development of new technology visual aids

Wildlife Hazard Mitigation ($16M)

- Biofuel and forage crop evaluation
- Directed energy evaluation
- Integration of avian radar information into ATC procedures

Airport Planning and Design ($12M)

- Continuous updates of simulation enhancements
- Development of 3-D animations

Airport Technology Research Taxiway ($20M)

- Engineering design
- Construction
- Installation of new technologies

Airport Technology Research Facilities

NextGen Pavement Materials Laboratory

Subgrade Storage & Processing Facility

Braking Research Aircraft

National Airport Pavement Test Facility

High Temperature Pavement Test Facility

Airport Technology Research Taxiway

FAA New Large Aircraft Fire Test Site

Airport Safety R&D Project Milestones

Fiscal Year	12	13	14	15	16	17	18	19	20	21
Winter Operations: ($23M) — Heated Pavements, Braking Performance										
Instrumented heated pavement evaluation at the NAPTF		▓	▓	▓	▓	▓				
Engineering study of a hybrid heated pavement system		▓	▓	▓	▓	▓	▓			
Development of brake system control in ASBS Simulation Laboratory		▓	▓							
Integration of developed brake control system into FAA Boeing 727			▓	▓						
Development of mathematical models that represent ASBS				▓	▓	▓				
Next Generation Aircraft Rescue and Firefighting: ($22M) — Cargo Aircraft, Advanced Composite Materials, 2nd Level Access										
Fire testing of cargo aircraft interior lining material	▓	▓	▓	▓						
Effectiveness of aircraft skin penetrating nozzles on interior cargo fires	▓	▓	▓	▓						
Development of guidance for tactics and agent requirements for cargo fires			▓	▓	▓	▓				
Evaluation of tools, tactics, and techniques for composite fires	▓	▓	▓	▓						
Hazards associated with composite fires; Investigation of Biofuels	▓	▓	▓	▓	▓					
Investigation of incidents involving interior fires	▓	▓	▓							
Development of design concept for interior access vehicle		▓	▓	▓	▓					
Visual Guidance: Safety, Capacity and Environment ($32M)										
Development of new heliport lighting		▓	▓	▓	▓	▓				
Identification of LCGS framework specifics		▓	▓	▓	▓					
Advanced Taxiway Guidance System Improvements		▓	▓	▓	▓	▓	▓	▓	▓	▓
Development of new technology visual aids		▓	▓	▓	▓	▓	▓	▓	▓	▓
Wildlife Hazard Mitigation ($16M)										
Biofuel and forage crop evaluation			▓	▓	▓	▓				
Directed energy evaluation					▓	▓	▓	▓		
Integration of avian radar information into ATC procedures				▓	▓	▓	▓			
Airport Planning and Design ($12M)										
Continuous updates of simulation enhancements		▓	▓	▓	▓	▓	▓	▓	▓	▓
Development of 3-D animations		▓	▓	▓	▓	▓	▓	▓	▓	▓
Airport Technology Research Taxiway ($20M)										
Engineering design	▓	▓								
Construction	▓	▓	▓	▓	▓					
Installation of new technologies					▓	▓	▓	▓	▓	▓

Airport Technology Research Facilities

NextGen Pavement Materials Laboratory

Subgrade Storage & Processing Facility

Braking Research Aircraft

National Airport Pavement Test Facility

High Temperature Pavement Test Facility

Airport Technology Research Taxiway

FAA New Large Aircraft Fire Test Site